rned
e b‹

How Did That Happen? –
Engineering Safety and Reliability

How Did That Happen? –
Engineering Safety and Reliability

by

William Wong
CEng, FIMechE, FIMarEST

**Professional
Engineering
Publishing**

Professional Engineering Publishing Limited
London and Bury St Edmunds, UK

First published 2002
Reprinted 2002

© W Wong

ISBN 1 86058 359 8

A CIP catalogue record for this book is available from the British Library.

Printed and bound in Great Britain by The Cromwell Press Limited, Wiltshire, UK.

Related Titles of Interest

For the full range of titles published by
Professional Engineering Publishing contact:

Sales Department, Professional Engineering Publishing Limited,
Northgate Avenue, Bury St Edmunds, Suffolk, IP32 6BW, UK
Tel: +44 (0)1284 724384
Fax: +44 (0)1284 718692
Website: www.pepublishing.com

This book is dedicated: -

To all engineers struggling to build a better and safer world.

Contents

Contents

Contents

Acknowledgements

This book could not have been written without help from:

Bechtel Ltd. For part time secondment to University College, London, so as to develop lectures on safety and reliability in design for 2nd year engineering students.

Dr Paul Williams, UCL Mechanical Engineering Department.

Prof. John Strutt Cranfield University, Chairman SRG, IMechE SRG committee members Andrew Bokor, Derek Heckle, and John Lee for their proof reading and comments.

J Harris (of Manchester University) for his notes on statistical analysis and RCM.

Clive Horrell, for his notes on noise.

Philip Highe, for his notes on radiation.

Roland Pruessner, GE Power Systems Essen, Germany, for providing examples of computer control screens.

The HSE for their comments.

Smit International for the photo of the Herald of Free Enterprise.

Prof. S Richardson Imperial College, for the photos (copyright unknown), and his notes on Piper Alpha.

Also to A Ogg, A Dooley, I Grey, C Bartley, D Wong, friends and neighbours for their proof reading and help in making it readable.

To all advertisers and others for their donations to help make the book affordable to students.

About the Author

William Wong is currently a visiting lecturer on safety and reliability at University College London.

William is retired after 25 years at Bechtel, holding senior positions in industry and working as a professional engineer for over half a century. His has worked on a wide range of projects; in the design and construction of North Sea platforms, petrochemical plants, power stations, gas and oil transmission pipelines, wind tunnel, and cryogenics. In his early years he was in manufacturing. He was in the aerospace industry on aero engine development and then in the oil industry on the design and manufacture of industrial gas turbines and process compressors.

Foreword

When major accidents occur with a large number of people injured or killed, people are shocked and ask 'how did that happen'? For every major accident there are hundreds more where only a few people are involved. These may not have a high profile, but they are just as shocking to the local community. Everyone wants to be safe from injury, accidental death or ill health due to pollution. There are laws and regulations that require this.

Very often the first impression following an accident is; that the people directly involved are to blame. Someone made a mistake. However, when all the facts are known, it is found that bad management and inadequate engineering are also to blame. For example, a badly trained inexperienced driver, driving a car with unreliable brakes, has a high probability of having an accident.

Managers and engineers need to learn about safety, health and environmental risk management. They need to understand the relationship between reliability, availability, maintainability and safety. They have to identify hazards and take action to avoid or reduce the risks that they pose. They have to understand their legal obligations and professional responsibilities. These subjects should be an important part of the education of engineers at all levels.

This book should be a useful textbook for undergraduates, engineers and others who need a comprehensive introduction to the basic principles of safety, health and risk management. Although the emphasis may vary they will be applicable across all industry sectors.

This book exceeds the recommended syllabus of the Hazards Forum, which is the Inter-Institutional Group on Heath and Safety, of the Institution of Chemical Engineers, Institution of Civil Engineers, Institution of Electrical Engineers and the Institution of Mechanical Engineers. The book also follows the recommendations of the Engineering and Technology Board.

Chapter 1

Background

1.1 Introduction

Engineers have been trained and educated in how to design and make things. Their focus has been on how to make things work. Safety has been considered as a part of Occupational Health and Safety, and in the past viewed as an administrative task associated with managing the work force.

As the twentieth century is left behind, more and more people live in a man-made environment. They depend on engineering and the application of science and technology for housing, electrical supplies, water supplies, the processing of sewage and refuse, transport, communications, the production of raw materials, and even the way food is produced.

The public takes for granted the reliability of all these services, and that all their needs will be supplied as and when required. Once in a while, however, people are shocked out of their complacency with an incident like Chernobyl, when a water-cooled nuclear reactor of Russian design overheated and caused radioactive material to be discharged to the atmosphere. This affected the whole of Europe causing an expected increase in the cancer death rate. It also affected the public acceptance of the use of nuclear power, even those of much safer designs, elsewhere.

The railway disasters that occurred in the UK during the years 1998–2000, with many dead and injured, also had an immediate effect. Disasters affect many more people than just the victims, see the memorial shown in Fig. 1.1. The work of the media has also focused public attention on safety and reliability, causing them to become political issues.

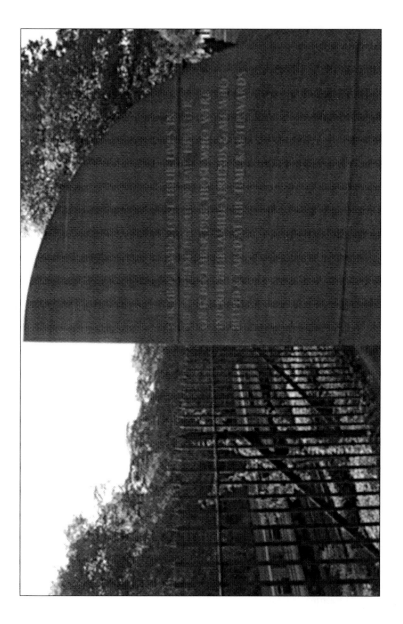

Fig. 1.1 Memorial to Clapham rail disaster

Examples of accidents are given in Table 1.1 **(1)**. All these accidents were the result of a failure, of men, machines and/or systems. In many instances the probability of such accidents could have been foreseen.

Table 1.1 **Man-made disasters**

Sector	Accident	Hazard	Fatalities	Year
Nuclear	Tokaimura, Japan	Radiation	None immediate	1999
	Chernobyl, USSR	Radiation	31 immediate	1986
Chemical plant	Mexico City, Mexico		>500	1984
	Bhopal, India	Poison gas	>2500	1984
	Flixborough, UK	Explosion	28	1974
Off-shore platforms	Sea Crest, Thailand		91	1989
	Piper Alpha, North Sea, UK	Fire	167	1988
	Alexandre Keilland, North Sea	Structural failure	123	1980
Aircraft	Nagoya Airport, Japan	Crash	264	1994
	Kenworth M1 motorway, UK	Crash	47	1989
	Concord, Paris, France	Fire due to ruptured fuel tank	113	2000
Shipping	Estonia 'ro-ro' ferry	Capsized	900	1994
	Herald of Free Enterprise	Capsized	188	1987
Railway	Oslo, Norway	Collision	35	2000
	Paddington, UK	Collision	31	1999
	Southall, UK	Collision	7	1998
	Kings Cross Underground, UK	Fire	31	1987

1.1.1   The environment

While this book is not about environmental issues, there is an overlap. The incident at Bhopal, given in Table 1.1, had reliability and safety issues but also had an impact on the environment. As will be seen later, all process plant and machines emit waste products to the environment. Engineers make a major contribution to the health and well-being of mankind. However, the very technical solutions they provide such as chemicals, transport and power generation deplete the earth's resources and cause environmental pollution. It is therefore important that engineers in their work ensure that they heed the need to conserve and enhance the environment.

In recognition of the importance of the need to preserve the environment, the World Federation of Engineering Organizations drafted a code. The European Federation of National Engineering Associations FEANI, and the United Nations Education Scientific and Cultural Organization UNESCO also approved the code. The code states:

'The responsibility of engineers for the sustained welfare of the community is an integral part of their professional responsibility. Therefore engineers should always:

- Observe the principles and practices of sustainable development and the need of future, as well as present, generations the world over.
- Assess and process the environmental impact of engineering projects from their inception, and then monitor to ensure that they will have minimum diverse effects on the environment, on the health and safety of all people, and on their social and cultural structures.
- Strive to accomplish their objectives with the lowest consumption of natural resources and energy, and the minimum production of waste and pollution of any kind.
- Maintain the importance of social and environmental factors to professional colleagues, employers and clients with whom they share responsibility and consult other professions about the potential negative impacts of their common endeavours.
- Foster environmental awareness within the professions and among the public, and promote environmentally sound working conditions.'

Further information can be found in reference (2). For the applicable regulations and their enforcement see references (3) and (4).

1.1.2 Reliability engineering, example of the lunar programme

The theories of reliability engineering were developed by the electronics industry and adopted by the aerospace industry as demonstrated by the successful mission to the Moon. Figure 1.2 shows the launch of a spacecraft.

Fig. 1.2 Spacecraft launch

The development programme to place man on the moon was a good example of reliability engineering and the application of *risk segregation*. The potential risk of failure was enormous. The crew could die in space due to medical problems. Could a safe landing and return be ensured? How could the problems of exploring the lunar landscape be foreseen and overcome?

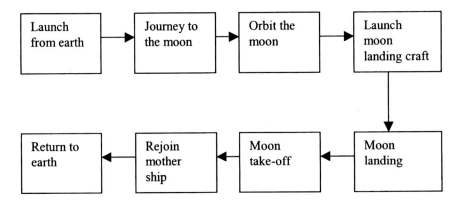

Fig. 1.3 The concept of the lunar space mission

Figure 1.3 shows the design concept of a moon landing as a block flow diagram. Each element can be developed in sequence. This is illustrated in the development programme shown in Table 1.2.

Even so, the programme was not without risk. During the ground testing of the Apollo spacecraft, a fire broke out in the three-man command module. Because of the pressurized pure-oxygen atmosphere inside the spacecraft, a flash fire engulfed and killed the three astronauts on board. As a result of this tragedy, the Apollo programme was delayed more than a year while vehicle design and materials underwent a major review. The review resulted in many failure modes being identified, more than if the accident had not happened. Major modifications to the original design had to be carried out. Following the initial successful lunar landing, more missions were needed in order to complete the gathering of scientific data required. Each mission provided more operating experience so that improvements could be made for each subsequent flight. These further missions are listed in Table 1.3. The accident that occurred on Apollo 13 illustrates the fact that 100 per cent safety and reliability can never be achieved. There is always an element of risk, which has to be managed. The good safety record of the Apollo programme owed a lot to the high calibre of the astronauts that were recruited and the close collaboration between them and the design engineers. The users (operators) will always think of more ways that things can go wrong than the designers. Risk can be minimized only when all parties, designers, builders, users, are involved in any given project/enterprise. The Apollo 13 mission returned to earth safely because the crew and their back-up team at mission control were able to put into practice a contingency plan which overcame the failure of the main oxygen supply. The risk to safety can only be minimized when everyone is conscious of the ever-present risk of failure and plans for it.

Table 1.2 The Apollo development programme

Year	Objective
1957/8	Launch small artificial unmanned satellites around the earth.
1959	Launch animals into orbit to study the effects of weightlessness and to verify the ability of mammals to live in space.
1960	Design and launch Mercury – manned spacecraft to test the ability to launch man into earth orbit and return safely.
1963	Successful completion of Mercury programme with six flights and up to three earth orbits achieved.
1964	Design and operation of Gemini spacecraft with a crew of two, to test the ability to operate for extended periods of time and to develop rendezvous and docking techniques with another earth-orbiting spacecraft.
1966	Successful completion of Gemini programme with ten flights and all required manoeuvres successfully carried out. One flight of 14 days provided the medical data needed for a lunar mission.
1967	Design and ground test Apollo spacecraft. Designed for a crew of three.
1968	First manned flight of a redesigned Apollo spacecraft, completing 163 earth orbits.
1968	First flight to the moon, completing ten lunar orbits and returning safely.
1968	Check out undocking of the lunar module landing craft from the main spacecraft, separate flight of each, and then rendezvous and redocking before returning to earth.
1968	Repeat manoeuvres with lunar module while orbiting the moon.
1969	First successful lunar landing and return to earth.

Table 1.3 The Apollo missions

Year	Mission	Purpose	Development
1969	Apollo 12	Gather more data from the same site as before	Improved guidance and landing techniques
1970	Apollo 13	To land on more rugged terrain	Mission aborted due to oxygen tank rupture
1971	Apollo 14	To complete the mission of Apollo 13	Improved oxygen system design
1971	Apollo 15	Explore mountain range	
1972	Apollo 16	Explore the highlands	
1972	Apollo 17	Explore a valley region	Last mission

1.1.3 Reliability engineering concepts

Reliability engineering as with any other technology has definitions of terms and units of measurement that need to be understood. A simplified analysis of the lunar programme can be used to illustrate these concepts.

Reliability

Reliability is the duration of failure-free operation under specified conditions (not misused). In predicting the reliability of a system or machine it is usual to observe the failure rate experienced. This is the number of times it fails in a given time. The expected reliability is then based on the probable duration of failure-free operation.

Failure

This is defined as a loss of function of a machine or system.

Fault

This is a non-conformance that needs unscheduled maintenance for correction, such as a car with a defective silencer.

Availability

Availability is the duration that a machine or system is able to function as required for the period under consideration.

Maintainability

As measured by the time it takes to repair a fault and return a system or machine to being available for operation.

Safety

To avoid the failures that could affect health or result in injury to people.

Probability

Nothing is certain in life, except death. In manufacturing it is impossible to ensure that all dimensions are the same. The principle of a tolerance has to be accepted with a probability of only some items having a specific dimension. Safety and reliability likewise cannot be predicted with certainty and only probable outcomes can be expected.

1.1.4 Analysis of the Apollo programme

Reliability analysis

- Number of attempts   7
- Number of failures   1
- Reliability is 6/7   86 per cent

Availability analysis

Assuming that the moon was in the correct position every month for a lunar mission.

- Period of lunar programme   60 months
- Number of attempts   7
- Availability is 7/60   12 per cent

Safety analysis

There were no fatalities or injuries and no reports of ill-health; 100 per cent safety was achieved.

Comment

The safety record was achieved at the expense of availability. Time was taken up in crew training, testing and preparation of the spacecraft.

1.2 The space shuttle disaster

Following the success of the lunar missions, the technology was then used to build and place objects in orbit around the earth for commercial and scientific use.

In spite of the success of the lunar programme, and after 14 successful missions, the Challenger II mission ended in disaster (5). This accident illustrates how reliability affected the safety of the crew. Risk is always present; failure can occur at any time.

The Challenger spacecraft had a component that was known to be unreliable. The political demands on its schedule, together with financial concerns, led to the risk of failure being ignored.

Any assembly of components will either work or not work. There are two possible outcomes and only one of them is acceptable. In order to improve the odds for success, a number of measures can be taken; for example, design modifications to make the consequences of failure more acceptable, testing to prove functionality before release to operations, and multiple endurance testing to establish repeatability. These are time consuming and expensive.

When under pressure to finish, there is a tendency to trust in luck and ignore the possibility of failure. It has been said that under such circumstances a lack of proof of unreliability is accepted instead of proof of safety.

To avoid disaster it is important to consider the consequences of failure. If people are able to focus on what would be the result of failure, then the pressures of money and schedule can be resisted, alternatives can be sought or the project abandoned.

Machines and systems, just like human beings, have an average expected life span. To predict the exact day on which something will cease to fulfil a function is impossible. Operating experience and the use of statistics, however, will enable a probable life span to be estimated. The use of monitoring techniques and the inspection of critical parts will give some warning of imminent failure.

1.3 Early steam locomotive explosions

The causes of industrial accidents are many and varied. An example, taken from the very early days of steam railways, will serve to illustrate this.

- Steam locomotives used to explode.
- Power output, hence the speed achieved going uphill, could be improved by stoking up the boiler to raise more steam pressure.
- To prevent excessive steam pressure, a simple steam safety relief valve was fitted. This consisted of a weighted lever, which held down a valve stopping the release of steam. When the steam pressure is excessive the force on the valve overcomes the effect of the weighted arm and steam is released to lower the pressure.
- At a steep hill, drivers got into the habit of tying a piece of string to the weighted arm. They pulled on it to stop steam being released. Inevitably a driver would cause the steam pressure to rise too high. The boiler would then explode causing loss of life.

The driver was blamed for causing the accident. In reality it is not that simple and the following questions need to be addressed.

- Was the allowed journey time too short?
- Did the locomotive need more power because the train was overloaded?
- Was the steam pressure relief valve releasing steam at the correct pressure?
- Did the locomotive lack power because it was in a poor state of repair?
- Was the wrong size of locomotive assigned to the train?
- Was the driver aware of the dangers arising from his interference with the steam relief valve?
- Why was the steam relief valve located within reach of the driver?

From the above it can be seen that management, engineering, operations and maintenance could equally be blamed.

Action to ensure safety and reliability therefore cannot be won for engineering reasons alone. Engineers will need to learn to make the case for action on much broader grounds. With major industrial accidents there will be a loss of capital, of revenue, as well as loss of life. In the case of the locomotive accident, for example, there will be the need to:

1. buy a new locomotive;
2. modify the design to prevent a repeat of the accident;
3. lose revenue until a new locomotive is available;
4. repair wagons;
5. repair the track;
6. clear the debris;
7. suspend operations while the track is cleared and repaired;
8. suffer possible loss of customers;
9. compensate for the loss of life.

Engineering projects have become increasingly complex and the engineer will need to learn how to deal with the following issues.

1. To reduce the risk to safety and reliability. This will require knowledge of how to identify hazards (dangerous things that could happen).
2. Once the hazards are known, what steps to take to try to avoid them.
3. To make contingency plans to minimize the effects if the worst should occur.

These general principles are applicable not only for safety, but also for a whole range of other situations as given in the examples below.

1.4 Examples of reliability engineering

Manufacturing
In the manufacture of high-value capital equipment, reliable delivery is important. Failure to deliver on time can have severe financial penalties. The manufacturing process can be split up into components and assemblies. The risk of delay due to mistakes and defects can then be assessed for each component and assembly. The consequences will be extra time to rework or replace defective material. This will enable contingency plans to be made for elements that pose the highest risk.

For example, if castings have a high risk of failure with a long lead-time to replace, the provision of spare castings could be a low cost compared to lost revenue and interest charges caused by delayed completion of plant.

Construction
The same approach can be used for ensuring construction schedules. For example, prefabrication of items may be preferred to site construction to avoid delays due to bad weather.

Costing
When applied to cost forecasting, the identification of items which have a risk of budget overrun will enable action to be taken to minimize such risk by focusing attention on the items that need closer management monitoring and control.

Product development
In the development of a new product, the components can be categorized in order of risk. The design can be ranked with components categorized as: proven in-house, proven by competitors and unproven. Action can then be taken to reduce the risk of failure by focusing on the areas of highest risk.

1.5 General precepts

The general precepts to be learnt concerning safety and reliability are listed below.

1. Nothing can be 100 per cent reliable and safe.
2. Human beings, one day, will invariably make a mistake.
3. Reliability cannot be predicted without statistical data; when no data are available the odds are unknown.
4. Statistics based on testing or other people's experience can only give guidance on the probability of failure.
5. The odds against failure can only be improved by adding redundancy and diversity. The use of two different methods to hold up trousers – belt *and* braces for example, provides a most reliable solution.
6. Making things safe and reliable costs money. Engineers will always need to cost the price of failure for comparison.
7. A safe and healthy working environment can only be achieved if the factors that affect safety and health are understood.

8. Because everything runs like clockwork, operators and management will always be lulled into a sense of security and may do something dangerous. Safety and reliability requires constant vigilance. Risk must be managed.
9. Operators may by-pass a safety system for some reason and think that the hazard will not occur. One day it will and disaster strikes. Even if an alternative safeguard is used this could result in an increased risk. Any such manoeuvre requires a full risk assessment.
10. A modification or a change in use of a system, or existing design, can lead to a higher risk of failure and a complete reassessment must be carried out. For example the use of high-speed trains on existing tracks, and signalling systems designed for slower trains, will result in increased risk of collision due to signals being passed and derailment due to excessive speed.

1.6 The role of codes and standards

For many products there are design codes and standards to which they are designed and manufactured. In general industry and consumer products British Standards play an important role. Safety and reliability are of paramount importance in the petroleum industry for example. The Petroleum Institute issues guide-lines concerning safe practice. The American Institute of Petroleum has codes of practice which are regularly updated and have controlled the design and application of equipment in the oil industry for decades.

Codes and standards are essential for the design and manufacture of established products where the safety and reliability hazards have been identified by experience over a long period. They provide requirements for:

- the responsibilities of the buyer/user in specifying the purpose and use of the equipment;
- the responsibility of the manufacturer to make known the allowable operating parameters;
- material selection;
- sizing criteria;
- design features to ensure reliable operation;
- safety provisions;
- third-party verification of design calculations where applicable;
- quality assurance (QA) and quality control (QC) requirements;
- acceptance testing requirements;

- inspection techniques to be used and the acceptance criteria;
- third-party inspection and independent certification of materials and work processes to ensure adherence to standards.

Safety and reliability issues, however, also need to be addressed for developing technology and in the design of plant, services, processes and their operation. The actions that are needed will be similar to those shown above for the design and operation of equipment. These actions will increasingly be required by government regulation.

1.7 Summary

The background to the need for engineers to focus on safety and reliability in their work has been explored, and some fundamental ideas on why accidents happen have been given. The general precepts should serve to provide a basic understanding of the issues of safety and reliability on which the following chapters will develop.

First, engineers will need to know the laws and regulations that have been enacted as a result of public concern for safety. These lay down regulations to improve safety on all aspects of engineering.

1.8 References

(1) Strutt, J.E. and **Lakey, J.R.A.** (1995) Education, training, and research in emergency planning and management, IMechE Conference, paper C507/009/95 (MEP Ltd), ISBN 0 85298 854 7.

(2) The Engineering and Technology Board (1994) *Guidelines on environmental issues*, www.engc.org.uk

(3) Environmental Agency, UK, www.environment-agency.gov.uk

(4) Environmental Protection Agency, USA, www.epa.gov

(5) Rodgers, W.P. (1986) Presidential commission on the space shuttle 'Challenger' incident, GPO, USA.

Chapter 2

The Law on Health and Safety

2.1 Introduction

Man-made disasters have caused governments to hold public enquiries and to issue regulations to improve safety, both for the public and workers. In 1972, Lord Robens in the UK issued a report on Health and Safety at Work. Two important conclusions were reached.

- 'The toll of death, injury, suffering and economic waste from accidents at work and occupational diseases remains unacceptably high. New hazards and problems are emerging. Apathy is the greatest single obstacle to progressive improvement: it can only be countered by an accumulation of deliberate pressures to stimulate more sustained attention to health and safety at work.
- The primary responsibility for doing something about the present levels of occupational accidents and disease lies with those who create the risks and those who work with them.'

The effects of disasters are not limited to the workers alone, but extend to the general public. As a result, there is a universal concern for ensuring the health and safety of workers and the general public who could be affected by their work. This has been reflected by the Occupational Health and Safety Act 1970 in the USA, the European Community laws and the UK Health and Safety at Work Act 1974, which has been updated to include the later EC laws. A review of the list of man-made disasters in Table 1.1 shows that, in spite of all these laws, disasters still occur worldwide. However, the intent of these laws is for the good and they need to be understood.

The law is based on two fundamental principles:

- the concept of a general duty of care for all persons, i.e. workers, operators, users, customers;
- goods and services provided must be fit for the purpose and not result in any danger to health and safety when used for the purpose intended.

It means that all those responsible must 'take all reasonable care to avoid creating any danger of death, injury or ill-health to any person...'. This is an important requirement, which bears on a wide spectrum of engineering activities. It applies to all phases of engineering work, including investigation, assessment, selection, planning, design, manufacture and construction, operation, maintenance and ultimate disposal. Any negligence, if proven, can lead to criminal charges.

2.1.1   Goods and services must be fit for purpose
This is a contractual issue and is subject to civil proceedings, which may result in the need for restitution in the way of monetary compensation. If the goods and services have an effect on health and safety, the criminal law will also apply. The supply of a machine that breaks down and causes injury could therefore be the subject of both criminal and civil proceedings.

2.1.2   The concept of all reasonable care
In the EU this is satisfied when the following prescribed actions and procedures are carried out:

- risk assessment – to identify hazards and the risk to health and safety;
- reducing the risk to As Low As Reasonably Practicable (ALARP);
- maintenance to ensure safety in operation and in working conditions;
- ensuring adequate training, supervision and the provision of information;
- action to measure, monitor and control.

The laws and regulations in the UK, Europe and the USA that most affect engineers are summarized in the sections that follow.

2.2 The Health and Safety at Work etc. Act 1974, UK

This is a summary and paraphrase of the law and some of its regulations. They should not be taken to be a substitute for a study of the Act and its

Regulations. Part I of the Act will be of major concern to engineers, especially Sections 1 to 9.

Section 1

An outline of the aims and intentions of the act, which is based on the fundamental point:

'*The primary responsibility for doing something about the present levels of occupational accidents and disease lies with those who create the risks and those who work with them.*'

Section 2

This concerns the obligations of employers to their employees.

2.1 To ensure, so far as reasonably practicable, the health, safety and welfare at work of all their employees.
2.2 To provide and maintain safe plant and equipment and ensure the safe handling and use of substances.
2.3 To provide a Health and Safety Policy Statement.
2.4 and 2.5 To appoint employee safety representatives.
2.6 To ensure consultation with safety representatives.
2.7 To appoint a safety committee.

Section 3

Obligations of employers to non-employees – outside contractors, visitors, and the general public – for their health and safety.

Section 4

Obligation to provide safe premises, without risk to health.

Section 5

Obligation to control emissions by the best practical means.

Section 6

Obligation of manufacturers, designers, importers and suppliers to provide products that will not affect the health and safety of users when used for the purpose intended.

Sections 7 and 8

The duty of employees, and others, to co-operate with the employer in ensuring health and safety.

Section 9

The responsibility of the employer to supply free any required safety equipment for use by employees or others.

Some examples

- To comply with the law, a tin of household paint will have: instructions on its use; instructions on the health and safety precautions required; what it should not be used for, e.g. not for consumption; and what has to be done if consumed, i.e. go to see a doctor immediately.
- A bus will need regular maintenance and inspection to ensure that the essential systems are in good working order. The driver has to be trained in the emergency procedures to be followed in the event of a fire or crash. The bus itself must have clearly marked escape routes, and facilities to open emergency exits and isolate fuel supplies.

2.3 Regulations

The Act is supported by a raft of regulations that cover every possible industry and work situation – there are hundreds of them. A small selection of those that every engineer will need to be familiar with is given below.

2.3.1   The Management of Health and Safety at Work Regulations 1999 (MHSWR)

The regulations, with their reference number, giving the general duties required of the employer are given below:

3. carry out a risk assessment;
4. principles of prevention (Schedule 1 below);
5. health and safety arrangements;
6. health surveillance;
7. health and safety assistance (the need to appoint a competent person to ensure compliance with fire regulations);
8. procedures for serious imminent danger and danger areas;
9. contact with external services (for first aid, emergency medical care and rescue work);
10. provide information to all workers;
11. the need to co-ordinate and co-operate with other employers on the same site with regard to fire regulations.

There are many other regulations that deal with the welfare and safety of different categories of workers, their duties, and the employer's responsibilities, etc.

Regulation 4, principles of prevention

Schedule 1

a) avoid risk;

b) evaluate risk that cannot be avoided;

c) combating risk at source;

d) adapt the work to the individual with regard to the work place, work equipment, choice of working methods... so as to minimize their effects on health;

e) adapt to technical progress;

f) replacing the dangerous by the non-dangerous or the less dangerous;

g) developing a coherent overall prevention policy, which covers technology, organization of work, work conditions, social relationships, and the influence of factors relating to the working environment;

h) giving appropriate instruction to employees.

2.3.2 Emergency planning, Control of Industrial Major Accident Hazards Regulations 1999 (CIMAH)

These are applicable to situations where there is a potential for a major accident as established by risk assessment and a safety case submission. The risk assessment will have identified major hazards and their possible occurrence. The safety case will have stated the actions taken by management to minimize the risk from the hazards; for example, training, supervision and institution of controls and procedures. The major elements of the emergency plan will stipulate the action needed to:

- raise the alarm;
- save life;
- contain the incident and prevent its escalation;
- marshal the external emergency services: police, fire brigade, etc.;
- ensure adequate training of individuals in all procedures by the staging of simulated emergencies.

2.3.3 The Provision and Use of Work Equipment Regulations 1998 (PUWER)

In summary the regulations require that equipment provided for use in the work place is:

- selected to be both safe and suitable for the task;
- maintained in a safe condition;
- inspected to ensure safety, with quality assurance records;

- only used by, and accessible to, qualified persons who have received adequate information, instruction and training;
- equipped with suitable safety measures such as controls, protective devices, markings and warnings signs, etc.;
- generally conforms to any other related health and safety regulations that are applicable to the place of work.

There are also specific requirements that concern mobile work equipment, power presses and miscellaneous other equipment. A conformity assessment may also be required.

2.3.4   The Reporting of Injuries, Diseases and Dangerous Occurrences Regulations 1995 (RIDDOR)

There is a legal duty to:

1. Notify the Health and Safety Executive (HSE) area office in the case of industrial accidents with an injury or a near miss;
2. Provide a written report on an accident report form within 10 days.

2.3.5   The Control of Substances Hazardous to Health Regulations 1994 (COSHH)

The steps required are listed below.

1. Identify the hazardous substances; assess the risks and who might be exposed to them.
2. Decide what precautions are needed to minimize the risk.
3. Prevent or adequately control the exposure of people who might be at risk.
4. Monitor control measures and ensure that they are used and maintained.
5. Monitor the exposure of people to dangerous substances if exposure limits are required to be enforced.
6. Carry out the health surveillance of anyone who is exposed to any substance that can be linked to any particular disease or adverse health effect.
7. Inform, train and supervise.

Note

Chemicals (Hazard, Information and Packaging for Supply) Regulations 1994 (CHIP) lists all such substances. Under the regulations they must be labelled as such and must be accompanied by Safety Data Sheets that identify hazards, preventative measures, and emergency and first aid measures.

2.3.6   *Supply of Machinery Safety Regulations 1992 (EEC Machinery Directive 89/392/EEC)*

All new machinery, either a one-off or for series production, must comply with the regulations. The regulations also apply to any machinery imported into the EU, new or secondhand, and also to refurbished or modified machinery where used for a different purpose, or where the performance is improved from its original level.

Summary of actions required by the regulations

1. A risk assessment must be carried out and the essential health and safety requirements met by good design and the provision of guards and safety devices.
2. Operating and maintenance instructions, listing required safety precautions
3. A responsible person must issue a declaration of conformity.
4. A 'CE' identification mark must be affixed.
5. The machine is in fact safe.
6. A Technical File must be drawn up and retained for 10 years.

Required Technical File contents (in two parts)

Part 1

1. The name and the address of the manufacturer and the identification of the product.
2. The list of harmonized (EU) standards complied with by the manufacturer, and/or the solutions adopted to satisfy the essential requirements for health and safety.
3. A description of the product.
4. Operating and maintenance instructions.
5. Overall plan of the product, if any.

Part 2

1. Test reports
2. QA manual
3. Plans
4. Description of the products and processes
5. Codes and standards adopted

2.3.7   Electromagnetic Compatibility (amendment) Regulations 1994

This requires that apparatus shall be so constructed that:

1. The electromagnetic disturbance it generates does not exceed a level allowing radio and telecommunications equipment and other relevant apparatus to operate as intended.
2. It has a level of intrinsic immunity, which is adequate to enable it to operate as intended when it is properly installed and maintained, and used for the purpose intended.

As an example: a programmable process control system must not be affected, or prevented from operating as intended, because of electro-magnetic interference from, say, a fluorescent light. Neither must its use cause any equipment to be affected by the emission of electromagnetic radiation. A technical file is required together with CE marking. The regulation requires either self-certification to a recognized code or standard, or external certification via a notified body such as BASEEFA/EECS (for definitions see Chapter 12).

Other similar regulations are:

- Low Voltage Electrical Safety Regulations 1989;
- Pressure Equipment Regulations 1999 *General duties relating to supply of pressure equipment and assemblies*;
- Simple Pressure Equipment (safety) (amendment) Regulations 1994;
- Electricity at Work Regulations 1989;
- Fire Precautions (workplace) (amendment) Regulations 1999.

2.3.8   The Equipment and Protective Systems Intended for Use in Potentially Explosive Atmospheres Regulations 1996 (S. I. 1996/192), (ATEX Directive 94/9/EC, reference (1)

This directive harmonizes the technical and legal requirements of such equipment and systems for use throughout the EU. Equipment includes electric motors, compressors, diesel engines, light fittings, control and communication devices, and monitoring and detection equipment (including the parts which are located outside the hazardous area).

In order to comply, equipment and systems are required to be to CEN or CENELEC standards 'or to meet the essential requirements' of the directive. Where approved standards are not available for mechanical equipment then they must comply with the essential requirements. These essential requirements fall into three groups, as follows.

Common requirements. Dealing with general issues, material selection, design and construction, potential ignition sources, external effects, safety devices and safety requirements.

Equipment requirements. Control of: potential ignition sources, surface temperatures, safe opening, dust ingress or egress.

Protective systems. Covering the need to ensure their reliability in preventing or minimizing the effect of any explosion.

A technical file is required, just as in the Machinery Directive, see Section 2.3.6 above.

Equipment categories

The essential requirements differ in accordance to the risk defined as follows.

Category 1. Where an explosive atmosphere is present for long periods.

Category 2. Where an explosive atmosphere is likely to occur during normal operation.

Category 3. Where an explosive atmosphere is only likely under abnormal circumstances.

Category M1. Mining equipment that can remain energized in the presence of an explosive atmosphere.

Category M2. Mining equipment that must be de-energized when an atmosphere exceeds the lower explosion limit.

Conformity assessment requirements

Category 2 and M2. Non-electrical equipment.

Category 3 equipment. Manufacturer's internal assessment. The technical file, except for category 3 items, must be deposited with a notified body.

Category 1 and M1. Protective systems.

Category 2 and M2. Electrical equipment and internal combustion engines.

EC type examination by a notified body.

Definition of a notified body

The notified body is responsible for carrying out testing and conformity assessment of the design and Product Verification (routine auditing) of subsequent manufacture or alternatively Production QA (auditing of the manufacturer's ISO 9002 quality control system) as applicable.

The notified body means a national certification authority appointed by the national government as notified to the EC. The notified body for the UK is the Electrical Equipment Certification Service, EECS, otherwise known as BASEEFA.

It should be noted that there is a reciprocal certification agreement between EECS and Factory Mutual of the USA.

2.3.9   Provisions to ensure the safety in the work place where potentially explosive atmospheres could be present (ATEX Directive 99/92)

This is due to be enforced throughout the European Community from 30th June 2003. From this date all new installations must comply and existing installations have up until 30th June 2006 to ensure compliance. In effect the IP code area classification rules have been officially adopted. The directive also extends the rules to cover explosive dust clouds so that Zone 20, which is for the presence of explosive dust clouds, will be the equivalent of Zone 0, which is for the presence of flammable gases. That is, prefix 2 is added for explosive dust clouds.

A new departure is that area classification rules are to be extended so that mechanical machines will need to be certified in the same way as electrical machines. This also brings the potential need for retrospective certification for mechanical equipment in flammable hazardous areas. This will also be needed where the electrical equipment has not been certified in accordance with the ATEX Directive 94/9 as given above.

The directive furthermore requires the employer to draw up and keep up to date an 'explosion protection document'. Ideally this must be done during the design phase of a plant and certainly prior to operating the plant. The purpose of the document is to demonstrate in particular that:

- explosion risks have been determined and assessed;
- adequate measures will be taken to attain the aims of the directive, which is to ensure a safe and healthy working environment;
- work areas are classified into zones as applicable;
- all work places and work equipment, including warning devices, are designed, operated and maintained with due regard for safety.

The document must be revised when the work place, work equipment, or organization of the work undergoes any significant changes, extensions or modifications.

2.3.10 Construction (Design and Management) (CDM) Regulations 1994

The CDM Regulations place duties on all those who can contribute to the health and safety of a construction project. Duties are placed upon clients, designers and contractors and the regulations create a new duty holder –

the planning supervisor. They also introduce new documents – health and safety plans and the health and safety file.

The regulations cover the design, installation, commissioning, maintenance, repair or removal of mechanical, electrical, gas, compressed air, hydraulic, telecommunication, computer or similar services which are normally part of a structure or a fixed manufacturing or process plant.

Duties of the client (on initiation of a project)
To ensure the following.

1. Financial provision is made and time is allowed for safety requirements in the initial planning of a project.
2. Prior notice is given to the nominated authority (Health and Safety Executive in the UK) of the project.
3. The site development requirements and the identification of any applicable hazards have been established. The project's conceptual design, for issue to the project team, is available.
4. A planning supervisor is appointed from the project design team.
5. A competent contractor is appointed as principal contractor.

Duties of the planning supervisor
To ensure compliance with all applicable safety regulations throughout all stages of the project. Produce a Safety Plan and maintain a Safety File.

Safety plan
The plan should, at all stages of the project, show how all hazards are to be identified and the risk to health and safety lowered to an acceptable level. It should include a list of scheduled activities and the procedures to be used from design through to construction and handover. Typical contents are:

1. Develop safety team organization and assignment of duties.
2. Lay down procedures to identify hazardous materials of construction and operation.
3. List documents for safety review.
4. List required safety drawings.
5. List required safety equipment and facilities.
6. Programme HAZOP (hazards in operation) and HAZAN (hazard analysis) meetings.
7. Classify and identify critical machines and processes.
8. Nominate machines for hazard assessment.
9. Verify adequate maintenance facilities for safe access.
10. Co-ordinate safety and constructability meetings between design and construction.

11. Ensure safety-centred maintenance plans are prepared.
12. Monitor adequate provision for safety activities with regard to time and cost.
13. Ensure adequate training and safety management provisions for construction.
14. Maintain records and prepare safety file.

Safety file

The safety file is required to contain a record of all the as-built design features, including all the information on risks to health and safety that could arise from operations and maintenance, and the maintenance tasks needed for safe operation. The file must be updated during construction and given to the client on handover.

Duties of the designer

The designer is required to identify any risks to health and safety in his design that could arise during construction, operation or maintenance either from the materials used or the facilities provided. His design must include all reasonable and practical features to avoid these risks in accordance with the principle of safety integration. He must:

1. Make clients aware of their duties under the regulations.
2. Give due regard, in his work, to avoid or minimize risks at an acceptable level to health and safety.
3. Provide adequate information, to those who need it, about the risks to health and safety of the design.
4. Co-operate with the planning supervisor and, where appropriate, other designers involved in the project.

Duties of the client (as operator)

The client must receive the safety file prior to handover of the completed project and keep it safe for future reference. The file must be consulted concerning any maintenance work or any subsequent alterations to the plant. This in effect overlaps with the PUWER regulations. See Section 2.3.3 above.

It should also be noted that any alteration, renovation or major maintenance work subsequently required will be subject to all the fore-going regulations.

2.3.11 *Other regulations, which are prescriptive*

Some regulations affect the design of operator interfaces, e.g. control rooms and control cabins or capsules. These include:

- Noise at Work Regulations 1989;
- Workplace (Health, Safety and Welfare) Regulations 1992;
- Vibration of Manually Operated Machines, etc.

2.4 Enforcement of the law

The Health and Safety Commission is responsible for promoting the objective of the Act and putting forward to government proposals for regulations under the act. The Health and Safety Executive (HSE) is responsible for enforcing the law via HSE inspectors stationed at area offices located throughout the UK. Deciding what is reasonable and practicable is subject to the discretion of the HSE. Inspectors will, as necessary:

- Offer information, advice, and support.
- Issue formal improvement notices.
- Issue prohibition notices where there is serious risk of injury.
- Make variations of licences or conditions or exemptions.
- Initiate criminal prosecutions of individuals, including company directors and managers. Where a death is involved, a charge of manslaughter, or corporate manslaughter, will be considered.

Enforcement under the act may also be carried out by: Local Authorities, Agency Authorities or Chief Officers of the Police, depending on the work activity concerned. A case then has to be prepared for prosecution and judgment by the courts. If convicted, the costs of prosecution can be recovered and penalties imposed.

2.4.1 *Penalties*

Lower Courts
Can impose the following penalties.
- For failure to comply with formal HSE notices, or court remedy order: a fine of up to £20 000, 6 months' imprisonment, or both.
- For breaches of Sections 2 to 6 of the Health and Safety at Work Act: a fine of up to £20 000.
- For other breaches: a fine of up to £5000.

Higher Courts

Can impose the following penalties.

- For failure to comply with formal HSE notices, or court remedy order: an unlimited fine, or up to 2 years' imprisonment, or both.
- For contravening licence requirements, or provisions relating to explosives: an unlimited fine, or up to 2 years' imprisonment, or both.
- For breaches of the Health and Safety at Work (HSW) Act, or of relevant statutory provisions under the Act: an unlimited fine.

Section 47 of the HSW Act provided that breach of the Act will not give rise to a civil action, but breach of any regulation made under the Act is actionable unless the regulations say otherwise as, for example, the Management of Health and Safety at Work regulations.

Recovery of damages

For workers and other parties to recover damages as a result of an accident requires considerable cost. Much ingenuity must be expended in the investigation, developing the pleadings, and the outcome of the trial can be uncertain. In general, successful actions have been based on the Tort of Negligence and/or the Tort of Breach of Statutory Duty.

2.5 Occupational Safety and Health Act (OSHA) 1970, USA

1 Purposes of the Act

To ensure safe and healthful working conditions for working men and women; by authorizing the enforcement of the standards developed under the act; by assisting and encouraging the States in their efforts to assure safe and healthful working conditions; by providing for research, information, education and training in the field of occupational safety and health; and for other purposes.

An abbreviated summary of part of the Act is given below. The full text is available from the OSHA website if required **(2)**, **(3)**.

2 Congressional findings and purpose

(a) This finds that the personal injuries and illnesses arising out of work situations are a substantial burden.

(b) 'to provide for general welfare, to assure... safe and healthful working conditions, to preserve human resources':

    (1) by encouraging employers and employees to reduce... safety and health hazards... to... provide safe and healthful working conditions;

(2)   by providing that employers and employees have separate but dependent responsibilities and rights…;

(3)   by authorizing the… setting of standards and the creation of a commission for adjudication;

(4)   by building on advances already made through existing initiatives;

(5)   by providing for research and development and innovation;

(6)   by exploring ways to discover latent diseases, establishing causal connections and other research into occupational health;

(7)   by providing medical criteria to assure that no one suffers diminished health, functional capacity or life expectancy due to work;

(8)   by providing training programmes…;

(9)   by providing… occupational health and safety standards;

(10) by providing an effective enforcement programme;

(11) by encouraging the States to assume responsibility…;

(12) by providing for appropriate reporting procedures…;

(13) by encouraging joint labour–management efforts….

3 Definitions

This section defines the meaning of the terms used in the Act.

4 Applicability of the Act

This confirms that the Act is applicable for the various States and regions of the USA.

5 Duties

This lays down the duties of the employer to provide a place of work free from hazards. Both the employer and employee have to comply with OSHA standards.

6 Occupational Safety and Health Standards

This provides the procedures under which OSHA standards can be established.

7 Advisory committee, administration

The establishment of a National Advisory Committee to advise on the administration of the Act.

8 Inspections, investigations and record-keeping

The authority to enter any work place, to inspect and to investigate. The right to require employers to keep records and to make periodic reports on work-related deaths, injuries and illnesses.

9 to 19 The legal procedures for enforcement
Section 13
This allows a court order for closure of a plant to counteract imminent dangers.
Section 17
This lists the penalties that can be imposed for each and every violation of a requirement or regulation. If a violation results in the death of an employee, the punishment can be a fine or imprisonment or both.

20   Research and related activities

21   Training and employee education

22   A National Institution for Occupational Safety and Health

23 to 34 Finance and other provisions for the administration of the Act

2.5.1   OSHA standards and the Federal Regulations

Standard industrial classification (SIC)
For the purposes of occupational health and safety, all work activities are divided into industries by a standard industrial classification (SIC) code. These divisions are:

A   Agriculture, forestry and fisheries
B   Mining
C   Construction
D   Manufacturing
E   Transportation, communications, electric, gas and sanitary services
F   Wholesale trade
G   Retail trade
H   Financial insurance and real estate
I   Services
J   Public administration

By the use of the SIC code and the number of employees located at a site, it is possible to obtain from the OSHA website a list of cited standards together with the penalties imposed for non-compliance in a previous year. This is a useful checklist (3).

Free consultation on the identification of hazards is also available from OSHA on application. The service is intended to improve occupational health and safety. It is confidential and any contraventions found will not be reported for enforcement, so long as the recommendations are carried out in a timely manner. This is mainly intended for small businesses and is independent of the OSHA inspection effort.

OSHA standards of particular interest for engineers together with highlights of their requirements are summarized below.

2.5.2 *Process Safety Management of Highly Hazardous Chemicals, OSHA standard, Federal Regulation 1910.119 (1992)*

This standard applies to any process that manufactures or uses toxic, reactive, flammable or explosive chemicals, including flammable liquids and gases. A process is defined as any system of interconnected vessels or even a separate vessel that could be involved with a hazardous release. The standard lists some 130 specific and reactive chemicals that are considered highly hazardous. It should be noted that the Clean Air Act Amendments (1990) is also concerned with the prevention of accidental releases of chemicals to the environment and some of their requirements have been incorporated **(4)**. To be in compliance with the OSHA regulation the following actions are required.

Process safety information

This is in effect the safety file that must be developed and maintained on record. It must contain written safety information identifying work place chemicals and process hazards, and equipment and technologies used in the processes. As a minimum it should also contain:

1. Material safety data sheets (see standard 1910.1200 chemical hazard communication).
2. Block flow diagrams.
3. Process chemistry.
4. Maximum intended inventory.
5. Safe upper and lower limits of operating parameters.
6. The consequences of deviation and their impact on health and safety.

Data on equipment should include:

1. Materials of construction.
2. Piping and instrumentation diagrams (P&IDs).
3. Electrical classification.
4. Relief system design and design basis.
5. Ventilation system design.
6. Design codes and standards used.
7. Material and energy balances.
8. Safety systems installed.

The file is to be used for hazard analysis, management of change, the conduct of audits and incident investigation.

Process hazard analysis
This is the process of hazard identification, risk assessment and control. One or more of the following methods must be used to determine the hazards and evaluate the risks of the process being analysed:

1. What-if.
2. Checklist.
3. What-if/checklist.
4. Hazard and operability (HAZOP) study.
5. Failure mode and effects analysis (FMEA).
6. Fault tree analysis.
7. An appropriate equivalent methodology.

The result of the analysis should result in the determination of the actions needed to reduce the risk of the hazard arising. These will include:

1. The identified hazards from the process.
2. The identification of any previous incidents that had a potential for disaster.
3. Applicable engineering and administrative controls of the hazards.
4. The consequences of the failure of the engineering and administrative controls.
5. Facility siting.
6. Human factors.
7. A qualitative evaluation of a range of the possible health and safety effects on employees in the work place if there is a failure of controls.

The employer is required to carry out the recommendations with a written record of the action taken and a timetable of their implementation. The required actions will involve operating and maintenance personnel, the safety department and others who will need to be informed and consulted.

The process hazard analysis must be revalidated and updated at least every 5 years by a team, as required by the standard's requirements, to ensure that the hazard analysis is consistent with the current process.

Operating procedures
Develop and implement written operating procedures for the chemical processes, including procedures for each operating phase, with operating limitations and health and safety considerations.

Employee participation

The employer must consult with employees and their representatives on the development and conduct of hazard assessments and the development of chemical accident prevention plans, and provide access to these and other records required under the standard.

Training

Initial training

Each and every new process operator must be trained in:

1. An overview of the process.
2. Its operating procedures.
3. The specific health and safety hazards of the process.
4. Emergency procedures and shut-down procedures.
5. Any other appropriate safe working practices.

Refresher training

This must be provided for each operator after initial training at an interval of not more than 3 years. Refresher training should be carried out more frequently if needed.

Training documentation

Maintain records of the date at which training was provided for each operator together with a verification of the effectiveness of the training given.

Contractors

The employer, his contractors and their employees have the same joint responsibilities for health and safety. The employer must provide the same information and training as applicable for his own employees.

Pre-start-up safety review

This must be carried out before any hazardous chemical is introduced to any new facility or modified facility, which requires a change in the process safety information file. The safety review must confirm the following:

1. Construction and equipment are in accordance with design specifications.
2. Safety, operating, maintenance, and emergency procedures are in place and are adequate.
3. A process hazard analysis has been performed and implemented, and the management of change requirements has been met.
4. The training of process operators has been completed.

Mechanical integrity
This applies to the following equipment:

1. Pressure vessels and storage tanks.
2. Piping systems (including piping components such as valves).
3. Relief and vent systems and devices.
4. Emergency shut-down systems.
5. Controls (including monitoring devices and sensors, alarms and interlocks).
6. Pumps.

QA documentation is required to ensure that the equipment is designed, manufactured and installed suitably for its proposed application. The employer must then establish and implement written procedures to maintain the ongoing integrity of the equipment. Records and QA documentation of all tests and inspections must be kept available for inspection as required.

Hot work permit
A hot work permit system must be implemented for any operations on or near a covered process. It must document compliance with OSHA regulations in accordance with standard 1910.252 (a).

Management of change
Before any change takes place, the following must be considered:

1. The technical basis for the change.
2. Impact on employee health and safety.
3. Modifications to the operating procedures.
4. Necessary time period for the change.
5. Authorization requirement for the change.

If after due consideration the change is decided upon, then all the requirements of this standard will need to be reapplied, starting with the revision of the process safety information.

Incident investigation
This must be carried out within 48 h of any incident that resulted in, or might have resulted in, a catastrophic release of a highly hazardous chemical in the work place. Reports must be kept for at least 5 years and they must include:

1. Date of the incident.
2. Date the investigation began.
3. Description of the incident.

4. Factors that contributed to the incident.
5. Recommendations resulting from the investigation.

The employer is required to produce an action plan, and the resolutions and corrective actions carried out must be documented and filed with the report. (Also note standard 1904.008 regarding the reporting of fatalities/multiple hospitalization injuries.)

Emergency planning and response
Compliance with the following standards is required:

- Employee emergency plans and fire prevention plans standard 1910 38.
- Hazardous waste operations and emergency response standard 1910 120.

In essence the following actions are required:

1 *Planning* to account for:
- Emergency escape.
- Procedures for an emergency team to shut down critical operations.
- Procedures to account for all employees following evacuation.
- Appointment and function of emergency rescue and first aid teams.
- Preferred reporting system for emergencies.
- Contact names for further information on emergency planning.

2 *Nomination of the emergency response team and team co-ordinator.*
3 *Training of the emergency response team for the different types of disasters to be expected.*
4 *Training of the employees:*

- Evacuation plans.
- Alarm systems.
- Reporting procedures for personnel.
- Shut-down procedures.
- Types of potential emergencies.

5 *Training must take place at random at least once a year.* Management and employees must evaluate the performance immediately. When possible a full-scale exercise should be held to include outside community services.

Compliance audits

Employers are required to certify that compliance with this regulation has been carried out at least every 3 years. A report of the findings, recording any deficiency found and corrective measures taken, must be written. A copy of the last two audit reports must remain on file, available for examination at any time.

2.5.3   Chemical hazard communication, OSHA standard Federal Regulation 1910.1200, reference (5)

This lays down the duties of chemical manufacturers, importers, distributors, employers and employees concerning the issue of material safety data sheets, safety procedures, precautions, instruction, information and training. Material data sheets must give information on:

1.   Chemical identity (of the material).
2.   Physical and chemical data.
3.   Fire and explosion hazard data.
4.   Reactivity data.
5.   Health hazard data.
6.   Preventative measures.
7.   Emergency and first aid measures.

These are similar to the COSHH regulations, see Section 2.3.5.

2.6 Summary

Engineers anxious to carry out their professional and moral duty to ensure the health and safety of workers and the public should find the regulations useful. Engineers will also find the regulatory authorities, HSE in the UK, and OSHA in the USA, helpful in providing further guidance in these matters. They publish an extensive series of guides, covering all industries. Advice on applicable regulations via a helpline is available and OSHA also provides training facilities for employers to attend. Much information is also available from their websites: see below **(3), (6)**.

It should be noted that other regulatory bodies control aircraft and ship safety. For example, shipping is controlled by the International Maritime Organization, as ratified by its member nations, and is regulated by codes covering all aspects of health, safety, and the environment. Of general interest will be the ISM (International Safety Management) code, as the basic principles are of universal application and generally mirror the requirements of other regulatory bodies **(7)**.

2.7 References

(1) Information sheets as available from the EECS.

(2) Occupational Health and Safety Act of 1970 with amendments.

(3) OHSA website, www.osha.gov/oshstats/std1.html

(4) OSHA 3132, Process safety management.

(5) OSHA 3084, Chemical hazard communication.

(6) HSE website, www.hse.gov.uk

(7) IMO website, www.imo.org

Chapter 3

Identifying Hazards

3.1 Introduction

The law requires a risk assessment to be carried out. The first step in a risk assessment is to identify hazards. These terms need defining:

- Hazard means anything that has a potential to cause harm (e.g. chemicals, fire, explosion, electricity, a hole in the ground, etc.).
- Risk is the chance, high or low, that someone will be harmed by the hazard.

All engineered machines and processes are potentially hazardous. They also give out emissions that can affect the surrounding environment and have an impact on health. It is important to know what the hazards are.

3.1.1 Hazards to humans

- Skin contacts by chemicals (acids, alkalis, etc.) which have an immediate destructive effect.
- Damage from petroleum products to skin properties – possible cancerous effects from long-term exposure.
- Penetration by sharp objects, by high-pressure jets – air penetration into the bloodstream can cause death.
- Inhaling polluted air.
- Eye contact by spray, mists, high vapour concentrations and harmful rays that can damage or destroy its tissues.
- Ingestion of contaminants. Taken through the mouth due to toxins entering the food chain or drinking water.

- Physical damage to the body.
- Loss of life support, e.g. temperature extremes, lack of oxygen.
- Effects on health can be immediate, or by long-term damage to body organs.

3.1.2   Hazards from machines and processes

The use of a block flow diagram such as Fig. 3.1 is a useful tool for auditing the presence of hazards in any machine or process stream.

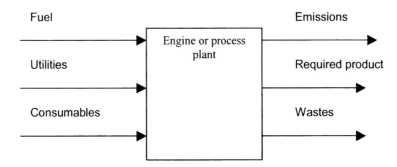

Fig. 3.1 Process stream block diagram

This is illustrated by some typical examples of hazards.

1. Gas turbines have noise, heat radiation and exhaust gas emissions that are a danger to health. The fuel could be inflammable natural gas. Other hazards are hot surfaces and moving parts. Special design features are needed to reduce the risk (e.g. silencing, guards, ventilation systems, fire and gas detection systems, features to reduce NOX emissions, etc.).
2. Airliners at 10 000 m are flying in an environment that cannot support human life. The passengers have to be protected from the hostile environment, and the noise, vibration and exhaust emissions from the engines. HVAC (heating, ventilation and air conditioning) systems have to be installed to maintain living conditions. The engines have to be fitted with silencers and antivibration devices. The safety of the aircraft depends on the reliability of its structure, control systems and its crew. The inherent risks are reduced by providing emergency oxygen supplies for each passenger in case of HVAC failure, redundancy in the aircraft structure and triple redundant control systems. Each aircraft has a pilot and co-pilot in case one becomes ill.
3. Chemical plants handle hazardous processes and contain toxic substances. A malfunction may lead to the release of fluids dangerous to life. To prevent this, plants have to be designed with the hazards

segregated. Control systems need to use the principles of redundancy and diversity to ensure reliable operation. They are operated by fail-safe automatic control systems; triple redundant detection systems may be needed for reliability; and emergency shut-down systems are required. Operators also have to be trained in emergency procedures.

The above examples also illustrate the principle of safety integration, which means that hazards are recognized during design and measures are taken either to eliminate or reduce risk.

Hazard identification is, therefore, an important issue for engineers. First will be the need to consider the hazards that affect the environment with an impact on health, and then the hazards to safety due to circumstances that result from operations or system failure.

3.2 Hazards from emissions

3.2.1 Waste emissions

All machines and engineered process plants produce waste streams; they are unwanted emissions. At the start of the industrial revolution, no thought was given to these emissions. It was assumed that the sky, the earth and the oceans were an infinite sink into which all manner of waste could be discharged with no harmful effect.

Due to the insatiable demand for energy, and the extravagant use of hydrocarbon fuels, the atmosphere now has a greater content of carbon dioxide. The earth can no longer absorb the CO_2 produced. In the hundred years following the industrial revolution, the CO_2 content of air has increased from 260 to 350 p.p.m. and it has been rising at the rate of 0.4 per cent per annum. CO_2 in the atmosphere reflects back infrared rays emitted by the earth. This is the greenhouse effect that contributes to global warming. Governments, and hence engineers, are now called to design plants to burn cleaner fuels with greater efficiency and to peg CO_2 emissions at the 1999 rate.

Emissions of gases heavier than air (including CO_2 and refrigerant gases) can collect at low points or confined spaces. This will displace the oxygen needed for life support.

Sulphur oxides and nitric oxides from the combustion of fuel and some industrial processes, when discharged into the atmosphere, combine with the water in the air and come down as acid rain. This causes the destruction of forests and aquatic life in lakes, and causes asthma.

Discharge of metals and inorganic materials can cause the pollution of waterways and oceans. Materials that do not biodegrade can quickly be taken up by marine life and, via the food chain, bioaccumulate in humans.

Water pollution

There are no specific limits given for pollutants that can be discharged with waste effluents. The limit is considered on a case-by-case basis by the local Environment Agency office. As an example, 5 ml/l oil contamination of wastewater is a typical figure. One doubtful practice, which is contrary to the spirit of all regulations, is to dilute the effluents with fresh water and so reduce the concentration. This does not reduce the quantity of contaminant involved and there is now evidence of toxic material accumulating in the arctic regions to the detriment of marine life. Water pollution effects are shown in Table 3.1.

Table 3.1 Water pollution effects

Pollutant	Effect
Oil	Generally biodegradable (but reduces the oxygen balance), fouling of birds, impact on reefs
Organics	PCBs, DDT, etc., chemical pesticides banned due to their bioaccumulation toxicity
Nutrients	Eutrophication, for example when lakes are enriched with nutrients, causing abnormal plant growth, excessive decay and sedimentation, and destruction of fish life
Metals	Cadmium, lead, mercury, copper, zinc. Bioaccumulation, rapid take-up by marine organisms, loss of marine foods, health impact

For example

A chemical plant on Tokyo Bay discharged effluent contaminated with mercury compounds. After a period of time, villagers living off the fish from the bay suffered mercury poisoning. This then attacked the brain and kidneys and affected their nervous systems. This is an example of bioaccumulation where toxic material is not degraded by biological action, is absorbed, accumulated and passed on from one species to another. The whole food chain becomes contaminated and affected.

Air pollution

In the case of air pollution, however, there are strict regulations on the amount of pollution and the period of exposure allowed to protect health. This is in addition to the actions needed to protect the environment (see Table 3.2). The allowable pollution is measured in mg/m^3. Normally emissions become diluted by dispersion into the atmosphere. Under freak weather conditions they can become concentrated, with disastrous results.

Table 3.2 Air emission effects and air quality regulations

Pollutant	Impact	Exposure	USA limit	EC limit
Sulphur dioxide from the combustion of sulphurous fuels	Effects on health, plant and aquatic life (acid rain)	24 h	365 $\mu g/m^3$	400 $\mu g/m^3$
		1 year	80 $\mu g/m^3$	140 $\mu g/m^3$
Hydrogen sulphide from processing of acidic gas, crude oil and paper pulp	Exposure to small concentrations will cause lung damage; higher concentrations will cause immediate death due to flooding of the lungs			
Nitrogen oxides (NOX) from combustion of fuels, nitric acid, explosives and fertilizer plants	Degenerates to nitric acid; affects health; in the presence of sunlight combines with hydrocarbons and causes photogenic fog, and contributes to global warming	24 h		200 $\mu g/m^3$
		1 year	100 $\mu g/m^3$	80 $\mu g/m^3$
Particulates, less than 10 μm size from industrial emissions	Lung disease, loss of immunity, property damage	24 h	150 $\mu g/m^3$	300 $\mu g/m^3$
		1 year	50 $\mu g/m^3$	150 $\mu g/m^3$
Carbon monoxide from incomplete combustion of fuels	Excessive exposure causes brain damage followed by death	1 h	40 mg/m^3	30 mg/m^3
		8 h	10 mg/m^3	10 mg/m^3
Carbon dioxide from the combustion of hydrocarbons	Global warming due to greenhouse effect; affects breathing rate	2–8h	Possible injury to health at concentrations over 5000 p.p.m.	
Organics	Ozone depletion, health impact and global warming	1 h	235 $\mu g/m^3$	
Heavy metals used in industrial processes	Especially lead, cadmium, arsenic; absorbed into the bloodstream through the lungs, they are bioaccumulators harmful to children	3 months	1.5 $\mu g/m^3$ of lead	
CFCs/halons	These are banned due to their effects on ozone depletion and hence global warming; it also results in increased ultraviolet radiation			

3.2.2　Heat emissions, ambient and surface temperature

Heat is emitted due to the inefficiency of industrial machines and processes. This may be discharged as waste hot water or hot air. Discharge into lakes or the sea will change the temperature at the point of discharge and so affect marine life. Engines and boilers heat the operating area where they are located and affect the operators in their vicinity.

Human beings must maintain their core body temperature within 35–38 °C. At lower body temperatures hypothermia occurs with loss of consciousness. Below 32 °C the heart will stop and death follows. At higher temperatures heat stroke occurs and, when the body reaches 41 °C, coma sets in and death follows.

Humans can live in environments higher and lower than the ideal body temperatures and the body will attempt to maintain its own temperature. People can survive for example in subzero temperatures. However, excessive exposure will cause loss of internal temperature control, with fatal results. In cases where workers are exposed to temperatures that exceed those normal to the location, exposure times will need to be monitored to ensure the health and safety of workers.

Hot surfaces at 49 °C and above, if touched, can cause skin damage and should be insulated. When surfaces are only subject to casual contact, such as within reach of walkways, it is common practice to only apply warning signs and/or personnel protection for temperatures of 65 °C and above. This is of course provided that there are no local regulations to the contrary.

3.2.3　Noise emissions

Engineers are not usually educated about noise yet their work causes noise pollution. Noise is an unwanted sound produced by working machinery and plant. The noise may be continuous, intermittent or erratic, depending on the source. It annoys, distracts and generally upsets and disturbs the tranquility of an otherwise peaceful environment. It can cause hearing damage. Noise also affects the ability to communicate, an important consideration in the design of control rooms and cabins.

The nature of noise

A pure sound is a pressure wave at a constant frequency. The sound pressure level (L_p) is measured in decibels (dB) and its frequency in Hertz (Hz). Machinery, however, produces noise that is an orchestration of many different sounds at different frequencies. An engine will produce sounds at different frequencies that are harmonics of the running speed made by its different components and the processes of combustion. Noise radiates outwards from its source and can be channelled to be directional. It is also reflected back from hard surfaces to cause an increase in noise levels. Absorbent surfaces will reduce this effect. Noise can be attenuated (reduced) by distance or by measures to dissipate its energy. A noise source in a container can be designed to be unheard outside. The amount of attenuation depends on the density of the wall and any noise absorptive materials used. Openings, which could allow the noise to escape, can be fitted with silencers that will absorb its energy and/or cause the noise to be reflected back inside.

Noise measurement

As a first approach a simple noise meter can be used to measure noise. This measures the noise in decibels (dB). The instrument usually has a number of scales, A, B and C. For most measurements the A scale, dB(A), is used. The human ear does not respond equally to all frequencies and so the scales are an attempt to allow a simple instrument to provide a single reading to represent what is heard. Weighting networks to discriminate against the low and high frequencies and adding the combined noise to give the overall level do this. Scale A gives the best approximation to the response of the human ear. However, precautions must be taken when using dB(A) levels. Different octave band combinations can produce the same dB(A) reading. Therefore an allowance has to be made if a pure tone component, or impulsive or irregular variations are present.

For a more accurate analysis of noise, an octave band analyser is used. This is used to measure the average noise level for the middle of each octave band. The frequency range is from 0 to 10 kHz divided into eight octave bands. Sometimes spectrum analysis may be necessary in order to identify a specific problem. This enables each individual sound to be measured for its sound pressure level in dB and its frequency.

Noise as a health hazard

Noise can cause hearing damage and is also a safety problem because it affects communication and can be a distraction. Hearing damage is a function of loudness and time of exposure **(1)**. It is generally agreed that a noise level of 85 dB(A) is the maximum noise limit for an 8-h shift. However, the rules for allowable exposure times for other noise levels differ between the EU and the USA.

The difference between USA and EU regulations

EU regulations assume that damage is proportional to the total energy to which the ear is exposed.

In the USA it is assumed that damage is related to the temporary loss of hearing of a young person. The allowable exposure of 4 h at 90 dB(A) is based on the requirement that it is interrupted by six quiet periods of a minimum of 5 min each at a noise level of 75 dB(A). It is based on the idea that intermittent noise, allowing time to recover, is less damaging than continuous noise.

In both the USA and the EU, if a person does work at levels above 85 dB(A) for the time period allowed, then following this exposure the person concerned must return to a quieter environment for the rest of the shift. The allowable exposure times for USA and EU regulations are shown in Table 3.3.

Table 3.3 Allowable noise exposure

Allowable exposure limit T h	USA noise level dB(A)	European noise action level dB(A)	Effect and/or action required
Nil	135	140	Instantaneous irreversible damage
2	95	91	Above 90 dB(A), hearing protection must be enforced; possible hearing damage may occur with 25% of people exposed for 30 years or more
4	90	88	In the EU a noise assessment is required above 85 dB(A) and hearing protection made available if requested
8	85	85	Commonly adopted as maximum level allowed for equipment
16	80	82	Negligible hearing damage risk in speech frequencies
32	75	79	At 75 dB(A) 97% of people will suffer no hearing loss, at all audible frequencies, after exposure for 40 years

As given in the table, workers should not be exposed to more than 85 dB(A) for more than 8 h as a norm. In other situations workers may need to work extended hours and the use of the following equation (which is the equation for the exposure times in Table 3.3) will give the maximum equivalent noise exposure to 85 dB(A) for 8 h.

$$L_{ep} = (10/n) \times \log_{10} \{1/8 \Sigma \{[C_1 \times 10^{(n\,L_{p1}/10)}] +$$
$$[C_2 \times 10^{(n\,L_{p2}/10)}] + (\text{etc.})\}\}$$

Where

    L_{ep} is the allowed normal noise exposure 85 dB(A)
    C_1, C_2 are the exposure times in hours
    L_{p1}, L_{p2} are the exposed noise levels in dB(A)
    n is a factor; use 1 for EU regulations and 0.6 for USA regulations

Example based on EU regulations for a 12-h shift
Find the maximum allowed noise level for a 12-h shift.

$85 = 10 \times \log_{10} (1/8 \times 12 \times 10^{(L_{p1}/10)})$
antilog $8.5 = 1.5 \times 10(L_{p1}/10)$
$316\,227\,766 = 1.5 \times 10(L_{p1}/10)$
$\log_{10} 316\,227\,766/1.5 = L_{p1}/10$
therefore L_p is 83.2 dB(A)

Workers on a 12-h shift should be restricted to a maximum noise level of 83.2 dB(A).

Worker working in process plant
The above equation can also be used for operators patrolling plant, passing through various noisy areas for differing time periods. However, it may be convenient to make up a table like Table 3.3 with the noise levels at each noise zone and the allowable exposure times as calculated from the equation. This then allows the use of the following formula:

    $C_1/T_1 + C_2/T_2 + C_3/T_3 ... = 1$

Where C_1 is the actual exposure time at the noise level in Table 3.3 and T_1 is the allowable exposure limit time, Table 3.3.

Example based on USA levels using Table 3.3
A person works 3 h at 90 dB(A) and 1 h at 85 dB(A). Find the maximum noise level allowed for the remaining 4 h of the working day.
    If exposure time C_1 is 3 h at 90 dB(A), and exposure limit T_1 is 4 h:

C_2 is 1 h at 85 dB(A) and exposure limit T_2 is 8 h

Then

$$3/4 + 1/8 + C_3/T_3 = 1$$

To solve, the required fraction C_3/T_3 has to be 1/8.

The person must work another 4 h at a noise level that allows him to be exposed for 32 h (because 4/32 = 1/8); from the table this is 75 dB(A). The worker must remain in an area at 75 dB(A) or less for the remainder of his shift.

Noise control

The best approach to noise control is by integration in the design of plant and machinery. For example, fan noise can be reduced by blade design; flow noise in pipework can be reduced by lowering its velocity or by noise insulation. Machinery vibration produces noise and this is increased by transmission to building structures. Good design of dynamic systems will reduce vibration and the isolation of machine vibration to prevent transmission will reduce noise. Machinery-generated noise from turbulence in fluid flow will radiate through casings and be carried out through connecting pipework. Most of this can be reduced.

When the required noise levels cannot be achieved then the next best thing is isolation into noise hazard zones where noisy equipment is separated from workers by noise enclosures, walls and by distance. By the use of isolating walls and insulated control rooms it is possible to isolate workers from noise during normal operation and even maintenance. Warning signs are then required to alert workers from entry into noisy areas without ear defenders.

Noise as a pollution hazard

In the design of plant any noise impingement into the neighbourhood is usually considered to be unacceptable pollution. At the start of any project it will be necessary to establish the ambient noise levels at the plant boundary and especially at all local inhabited areas. Typical rural noise levels away from roads are: average cottage, 50 dB(A); night-time, 40 dB(A). The actual measured figures will establish the design noise levels for the plant, which must of course be less. It is usually advisable to appoint a noise consultant to oversee the work through the design period and to verify the outcome. It is of interest to note that in one case the presence of a low-frequency noise was overlooked. This was inaudible but caused the cups and saucers and roof tiles to rattle at a distant cottage. It is difficult to attenuate low-frequency noise.

Reducing noise and vibration levels is of prime concern in the design of ships and offshore oil and gas facilities. This is due to the concentration of high-powered machinery in a confined structure. The health and safety of humans is regulated by the IMO code on noise levels on board ships. Research has shown that the noise transmitted into the sea also affects the marine environment. Noises produced by machinery on ships and by cavitation from ships' propellers generate low-frequency noise in the sea. Measurements have shown that ship-generated noise in busy shipping lanes can reach 90 dB(A) at 500 Hz. Low-frequency noises affect dolphins and whales, since they communicate with each other at these frequency bands. Excessive noise can damage their ability to hear, and it has been suggested that physical damage could be caused to lung tissue. Ears could be ruptured, resulting in haemorrhages. It has been said that a deaf whale might just as well be a dead whale. Based on present research findings, the IMO regulations are likely to be extended to protect marine life, especially in sensitive areas such as Alaska, Hawaii and the Arctic – areas frequented by the beluga and humpback whales.

3.2.4   Radiation emission
People need to be protected from radiation emissions. These notes give the consequences of excessive exposure and underline the need to enforce safety procedures and provide adequate design measures for shielding.

Light radiation
Many work processes and plant emit infrared (IR) and ultraviolet (UV) light. Infrared light will cause damage by heating, with possible loss of sight. UV light will cause tissue damage, particularly to the skin, and is linked to various types of skin cancer. It can also cause loss of sight.

Heat radiation
This is normally limited to 1.5 kW/m^2; higher rates need safety measures for personnel protection or better design to limit the radiation.

Non-ionizing radiation
This is the radio frequency (RF) radiation and electrical field emitted by equipment such as radio transmitters, radar installations, mobile telephones, microwave ovens and overhead high-voltage power cables. High levels of this type of radiation will heat the affected tissues, causing immediate damage (especially to brain tissue) and even death. However, the effect of lower levels (such as emitted by mobile telephones) is not fully understood, although long-term exposure has been linked to certain forms of cancer and memory loss.

Ionizing radiation

This is the radiation emitted by radioactive equipment and materials, such as:

- naturally occurring radioisotopes, e.g. uranium ore and radon gas;
- operating nuclear reactors, which emit high-intensity gamma rays and high-energy (fast) neutrons;
- purified and man-made isotopes, e.g. nuclear fuel and nuclear weapons material;
- spent nuclear fuel and associated waste;
- research/scientific equipment, and medical radiation treatment equipment;
- X-ray machines.

The radiation is in two forms: electromagnetic and particulate.

Electromagnetic

Gamma (γ) rays (emitted by nuclear reactors and by some radioactive decay) and X-rays are extremely-high-frequency electromagnetic waves which are very penetrating and can cause very significant cell damage, leading to burns, cancers and immediate death.

Nuclear particle emissions

These are emitted at various energies; the emissions commonly encountered are listed below.

- Alpha (α) particles, which are helium nuclei, and therefore relatively large and slow with a very short range in air. These will only cause cell damage if ingested or if there is contact with the skin (burns).
- Beta (β) particles, which are electrons. These have a range of a few centimetres in air and will only cause cell damage if ingested or if there is contact with the skin (burns).
- Neutrons (n). These are emitted at very high energies by nuclear reactors and in radioactive decay. They have a very long range in air and can cause significant damage to human tissue, including burns and cancers.

3.3 Hazards from circumstances

3.3.1 The effect of elevation

Table 3.4 ICAO standard altitude table (extract)

Altitude (m)	Altitude (ft)	Temperature °C	Pressure (bars)
0	0	15	1.013
500	1640	11.8	0.954
1000	3281	8.5	0.898
1500	4921	5.3	0.845
3000	9843	−4.5	0.701
6000	19 685	−24	0.471
10 000	32 808	−50	0.264

To climb Mount Everest, which is at 9000 m, oxygen is needed to breathe, and protection is needed from the cold at −44 °C. Humans can usually live at altitudes up to about 1500 m. At about this altitude, the partial pressure of oxygen will have decreased to 0.179 bar (130 mmHg). Table 3.4 shows how air pressure and temperature change with altitude. The ability of oxygen to pass through the lung membrane will be reduced and performance is affected. HVAC measures will be needed.

3.3.2 Physical hazards

Latent energies are hazards, which if released could pose danger to life and limb. They can be categorized as follows.

- Potential energy release, such as people or loads falling from a height, due to failure of safeguards, restraints, structures and devices.
- Kinetic energy release, from explosions, release of moving components, due to failure of, for example, pressure vessels, components of engines and vehicles. Contact with moving parts. Impingement of high-pressure jets (can penetrate the skin and cause air to enter the bloodstream which results in death). Impact from loss of control of a high-speed train, aircraft or other vehicles.
- Electrical energy, due to failure of insulation, failure of safety procedures.
- Chemical energy – acid attack destroys skin and tissue.

- Fire, which is the most common form of chemical reaction, can cause immense loss of life and property.
- Radiation energy release.
- Consequential damage which indirectly affects other plant.

All these hazards are subject to statutory regulation and this checklist can be used to verify if they are present in any work process under examination. Outlines of some of the common hazards, including some that are often overlooked or disregarded, are given below.

3.3.3   Hazards due to vibration energy
All machines vibrate to some degree. As a result, noise emissions are produced, as discussed in Section 3.2.3. Vibration is also transmitted through structures.

Vibration affects the nervous system of humans. The use of hand-held equipment that vibrates can lead to hand–arm injury. The HSE gives guidance on practical ways to reduce risk.

3.3.4   Hazards due to electrical energy
It would seem that everyone is aware of the dangers of electricity but accidents still happen. Dangers can occur due to live components, insulation problems, fault conditions or residual stored energy. Electrical engineers are well trained in knowing the hazards, as are qualified electricians, and they must be consulted to ensure that all hazards are identified and the appropriate measures taken to minimize any risk.

3.3.5   Hazards due to chemical energy
Consulting the COSHH regulations can identify hazardous chemicals. These are all listed in the regulations. Manufacturers must affix warning labels and supply safety data sheets. These can be used to determine the hazards involved for the user. There are regulations concerning storage and the need for segregation into chemically compatible groups.

3.3.6   Fire hazard
This is the most common hazard, which is present in all areas of life. Most combustible materials are stored in a normal atmosphere, which contains oxygen, and so the risk of fire is then due to the possibility of an ignition source (see Fig. 3.2).

Combustible liquids can vaporize and so form an oxygen–air mixture at their surface that can be ignited. The temperature at which a liquid fuel vapour can ignite is called its flash point.

Ignition   Fuel

FIRE

Oxidant

Fig. 3.2 The elements needed for a fire

The heat needed for combustion to take place depends on the flash point if it is a liquid. Solids need a much higher temperature to ignite.

In the storage of materials it is usual to apply segregation according to their ease of combustion. This will ensure that if a fire starts in one place, it will not spread to another. The burning of plastics, for example, will cause them to liquefy and flow, causing rapid spread of the fire. The hazard of any fire is its rapid propagation, which will occur if there is inadequate separation and isolation of all combustibles in the vicinity.

Example of fuel segregation
Fire protection for boilers and engines must include automatic shut-off of fuel supply lines. The fuel tanks should be segregated by firewalls, or located at a safe distance away.

Example of oxygen fire
Atmospheres that contain a high proportion of oxygen will require less heat for ignition than a normal atmosphere. A steel pipe carrying pure oxygen can ignite and burn, just from the kinetic energy given up, say, due to a welding bead striking a bend in the pipe. A fire fed with oxygen will be fierce and intense, with burning metal. Oxygen is a serious hazard.

A patient suffered severe burns due to a fire started by his being resuscitated while being given oxygen. The staff did not know that the tiny amount of energy available from an electric spark was sufficient to start a fire when in the presence of oxygen.

Other hazards from fires
It is usual to fight fires with water so as to remove the heat required for combustion (see Chapter 5 for applicable technologies). The alternative method to extinguish a fire is by oxygen depletion. This can be used in enclosed areas, especially in rooms for electrical apparatus, where the use

of water could cause electrocution. In the event of fire, the room is sealed off from air and CO_2 is injected. This in itself also represents a hazard to any personnel present. Excessive concentrations of CO_2 can cause brain damage or death and there must be safeguards to avoid personnel being exposed.

The side effects of a fire also represent a hazard. Firstly the fire will deplete oxygen from the surrounding atmosphere. Most casualties from a fire die from the smoke and lack of oxygen. Secondly, especially where plastics are being burnt, the fumes could be toxic, and anyone exposed could die. *Evacuation is the first priority with any fire.*

The heating effect from a fire also causes other hazards. Liquids will expand and so increase in pressure if they are restrained in pipework or containers. This will also happen with gases. Liquid gas will boil when heated. On the other hand, metals when heated will become weaker unless they are of some special alloy. A fire can therefore result in explosions unless gases are released. Heat from a fire can also cause seals to become ineffective. Depending on the contents, the resulting leakage can present a further hazard.

3.3.7   The hazard of entrapment
In any abnormal situation, the normal means of access and egress could well be barred or congested, so that persons cannot escape. Situations involving fire, gas release or explosion could give rise to this danger. During the design phase, careful thought has to be given to this and the means of escape in at least two directions must be provided. Hence buses, for example, will have knock-out windows to allow escape in case the normal exit is blocked.

3.3.8   The hazard of entry
Entry into any container, tank, unventilated area or pit is a hazard due to the possibility that the atmosphere is toxic or lacking in oxygen. In other cases it may be hazardous because of restricted airflow, confinement or restricted access. People could faint or be entrapped. In all cases access must be controlled and unauthorized entry prevented. Monitoring of the atmosphere before entry and during work inside must be carried out. Constant communication with those inside has to be maintained from outside so that help can be summoned if rescue is needed.

3.3.9   The hazard of transfer operations
Any filling or emptying of any materials used in an industrial process has the danger of spillage and contamination of the people involved. The consequences will depend on the material. In the case of hazardous

chemicals, safety regulations will be involved. There are dangers even with non-hazardous materials such as filling or removal of lubricating oil. Spillage will cause a slippery surface, with a danger of people slipping.

3.3.10 The hazard of maintenance operations
The hazard is due to the possibility that all energy inputs have not been correctly dissipated, isolated and inhibited, e.g. fuel, electricity, utility feeds, pressurized systems, possible movement, presence of chemicals, etc.

3.3.11 The hazard of uncompleted work
Whenever work that is incomplete is left for another day or another shift to complete, there is a potential hazard. There is a danger of mis-understanding, which must be guarded against by proper communication. For example, a drain valve could be opened for draining a system prior to refilling. The next shift coming on duty, seeing that the system is empty and thinking that the system was ready for refilling, will open the refilling line and so lose the whole inventory. With proper communication the next shift would know that draining had not been completed and that the drain valve had yet to be closed. Misunderstanding can lead to disaster.

3.3.12 The hazard of change
This is one of the most serious hazards that is commonly overlooked. Any machine or plant that is operating reliably and safely could become very dangerous and unreliable if there is a change in use. A change in use could be in its function. A change in its relationship is often overlooked, such as having to work in conjunction with modifications. There may be a change to its design.

Any of these will have an impact on the way the machine has to work. In this situation there has to be a complete reassessment of its safety and reliability. There could be an interaction that has been overlooked; a component could be working beyond its capacity or capability; and safety factors could have been exceeded. Increase in use could cause fatigue limits to be exceeded. A complete review of its safety case or risk assessment will be needed, as will its maintenance and operating procedures. Operator retraining will also be involved and operating procedures will need revision. Emergency procedures could be affected.

3.3.13 The hazard of human error
Humans make mistakes. Everything must be done to reduce the risk of human error, which is forever present. The risk of human error has to be reduced by the formalization of work practices rigidly enforced and the provision of engineering controls. The following examples illustrate this.

3.4 A pump test facility accident

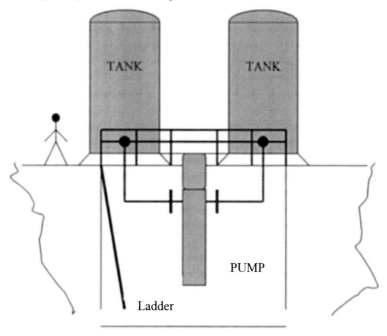

Fig. 3.3 The pump test facility

The test facility consisted of liquefied natural gas (LNG) tanks located adjacent to a pit, in which vertical multi-stage pumps could be installed for test. The installation had been in use for some time without incident.

On the day of the accident, an LNG pump was being subjected to a 24-h proof test. After running without problems during the day, it was left to continue running during the night, attended by two test observers. In the morning they were found dead at the bottom of the test pit. They died due to lack of oxygen. There was no requirement for the observers to go into the pit and it was thought that one had entered to pick something up and his friend went after him when he collapsed.

In the design of the test facility, the danger of falling into the pit was recognized, and the pit was safeguarded with railings. A steel ladder was provided to access the bottom of the pit. This was required during the installation of a pump for the test.

Any leakage of LNG will flash off into gas in the atmosphere. At first the gas will be cold at its boiling point of −160 °C and it will be heavier than air. As it warms up to ambient temperature it will become lighter than air and it then could become displaced by air. The cold methane gas was not completely displaced and the amount of air in the pit was not sufficient to support life. Any atmosphere with even only 60 per cent of the normal oxygen content could cause a person to faint.

Following the accident the fence was raised from waist high to man height and a locked gate was placed at the opening where the ladder was located. Warning signs were added. Air breathing apparatus and safety ropes were provided nearby as a means of rescue.

The facility could have been designed with no pit to avoid gas accumulation. This would have been more costly due to the need to raise the foundations of the LNG storage tanks.

This example demonstrates the need for:

- security to prevent uncontrolled entry;
- procedures to control entry;
- testing that an area is safe to enter;
- someone outside to monitor the situation and the person who enters;
- not attempting rescue alone but to summon help;
- rescue apparatus and a trained rescue team.

3.5 Needle stick injuries

Hypodermic syringes are used in medicine for taking blood samples and to inject drugs. After use, if the needle pricks another person there is a danger of transmitting a blood-borne illness. The needle is disposed of after use and therefore there is also the risk that someone could get injured in its subsequent removal. Even though all medical staff are aware of the dangers, injuries occur and so the syringes have had to be redesigned. Modern syringes are in two parts: the plunger capsule and the needle module. When required for use the two are mated. The needle has a sleeve which has to be removed before use. After use the needle module is separated into a disposal container that latches on to it and allows the plunger capsule to be extracted. There are other designs but they all follow the same principle of sheathing the needle until required for use and disposing after use into a special container.

Use of special disposal containers prevents subsequent risk of injury. The risk remains in its use. Inquiries have shown that some medical staff admit to suffering three to four injuries a year in spite of the special design features. Of course only a small percentage of patients suffer from AIDS (HIV), but there is a high risk from Hepatitis C and B. Medical staff need to be focused on the dangers involved; this is the function of risk management, which will be discussed in a later chapter.

3.6 Hazard analysis or risk assessment

The foregoing should enable the identification of the common hazards that can be present in the work place and in work processes. To assess risks for process operations, hazard analysis of the work place and work processes will be required. HAZAN, as it is sometimes known, consists of two stages. The first is to identify all the hazards present and the second is to consider the risk of it occurring and the consequences. This procedure is also known as risk assessment. In most situations where only simple routine operations are involved, risk assessment can be by one or other of the simple procedures given below.

Risk assessment (OSHA 'what if') procedure

1. Look for hazards
   - Make a list of all manual work or processes required from raw material to finished product.
   - Identify any hazards that may be present and the regulations that may be applicable.
2. Decide who might be harmed and how
   - Consider what risk the hazards may pose and to whom (not forgetting temporary contractors, visitors, persons with specific vulnerabilities, the public, etc.) and consider how the hazard, can be removed or minimized.
   - Consider what could go wrong if a mistake is made in a procedure or a fault develops in a process.
3. Evaluate the risks and decide whether existing precautions are adequate or if more should be done.
   - Evaluate the consequences of the mishap.
   - Consider what procedural or design safeguards are required by any applicable regulation or are needed to prevent, or minimize, any risk to health and safety.
4. Keep a record of the findings and produce an action list for implementation.
5. Carry out a regular review of work processes and revise the assessment if new hazards are found to be present.

Checklist procedure

Where the process is more involved, it can be split up into process units. Each unit can then be investigated and studied in all its aspects by a committee of experts on a prepared checklist of the following subjects.

1. Operator practices and understanding of the process.
2. Suitability of the equipment and the materials of construction used.
3. The chemistry of the process and the effectiveness of the control system employed.
4. Review of operating procedures and operating records.
5. Review of maintenance operations and maintenance records.
6. Make a list of 'what if' – what could go wrong for any of the above items which impact on health and safety.
7. Recommend actions to safeguard health and safety.

Other procedures

The above procedures provide a qualitative risk assessment that depends on the judgement of the audit team. Hazard analysis of complex processes and machine assemblies requires more sophisticated procedures. These will be discussed in later chapters.

3.7 Summary

Engineers are trained to be systematic and these are the same skills required for hazard identification. The different types of hazards explained in the foregoing should provide a good understanding of how they may be recognized. An introduction to the use of process block diagrams and HAZAN has been given. These provide a systematic way of identifying the presence of hazards so that action can be taken to protect people from the risk. It has also been shown that human error is an important hazard. It is clear that people are fallible and must be protected. It must be assumed that people will do stupid things and that things will go wrong.

In the design of plant, much thought has to be given to how humans behave. This is so important that the whole of the next chapter will be devoted to outlining the limitations of humans and the provisions needed to reduce human error.

3.8 Reference

(1) HSE publication IDG 75 *Introduction to the noise at work regulations.*

Chapter 4

Human Factors

4.1 Introduction

Human beings are involved with the operation of any plant or machine. Human beings make mistakes and when the consequences of these errors could lead to disaster there is a need to consider how these errors might be avoided.

Too often, when an accident happens, as in a railway disaster, the first thought is to call it driver error. The truth may well be much more complex.

The human–machine interface (see Fig. 4.1) has to be very carefully designed and this must be included in the design specification. Furthermore, the range of human types and operating environments must be clearly defined and considered.

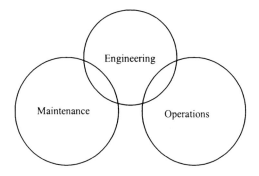

Fig. 4.1 Human interfaces to be considered

To ensure its safety and reliability, any plant or machine must be engineered so it can be operated correctly. Controls that are awkward, or instruments that give confusing signals, lead to error and a risk to safety. Equipment needs to be maintained to ensure safety and reliability. If clear instruction is not given on maintenance requirements critical to safety, then these will be neglected. If maintenance is difficult due to access or disassembly problems, then downtime is extended with a reduction in availability. Operator and maintenance participation in the design at an early stage is vital to safety and reliability.

Once designed and built, plant and machines will have to be operated and maintained by humans. It is also important that the range of people that will be involved is considered. People will vary in their education and physique, depending on where in the world they are located. These factors must be taken into consideration.

4.2 Ergonomics

This is the original label for human factor engineering. It is the idea that the design of machines and equipment should match the capacities of the people who will be the operators. The design must provide conditions that enable the people to function at their best. Their physical and mental limitations have to be considered, based on the type of people who will be involved.

4.2.1 The working environment

To ensure the safe operation of plant and machinery, working conditions must be provided to enable the operators to function efficiently without distraction. Typical design conditions for an engineered working environment are given in Table 4.1. They are for work areas located in a region with ambient temperatures of 33 °C in summer and –31 °C in winter. These will vary in accordance with local conditions, regulations and specified codes. They may also need to be adjusted to allow for other factors, as given in the notes.

The local ambient temperature will also affect the ideal working temperature. In the UK, for example, where the ambient temperature range is much lower, the minimum temperature for offices and control rooms is 16 °C, the maximum being dependent on what is comfortable. Adequate lighting must also be provided in accordance with the IES Code or other applicable regulations.

Table 4.1 Typical indoor design requirements for a location with local ambient temperatures ranging from 33 to –31 °C

Description	Air temperature (°C drybulb)	Relative humidity (% RH)	ACH-1 (Note 2)	Notional air pressure (Note 1)	NR sound (Note 3)	5–100 Hz peak velocity (mm/s)
Control rooms	22±2	50±10	6	+ve	40–50	1.5–2.5
Pump /compressor houses with potentially hazardous gases	40 max. 5 min.		12	+ve (Normal)	85 dB(A) max.	
Utility buildings with non-hazardous materials	45 max. 5 min.		6	–ve	85 dB(A) max.	
Chemical and additives stores	40/45 max. 5 min.		10	–ve	85 dB(A) max.	
Electrical substations and rack rooms	30 max. 10 min.	50±20	6	+ve	70	
Battery rooms	30 max. 10 min.		15	–ve	70	
Plant rooms	45 max. 10 min.		US	–ve	75	
Boiler rooms	45 max. 10 min.		Calculate	0/–ve	75	
Maintenance rooms	15 min.		10	0/–ve	55	
Gas turbine	10 min.		10	0/–ve	85 dB(A) max.	
Generator room	10 min.		10	0/–ve	85 dB(A) max.	
Offices	22±2	50±10	6	+ve	40	2.0
Workshops	15 min.		10	0/–ve	55	

Notes

1. Building room internal pressure relative to ambient may need to be adjusted for the following reasons:
    (a) +ve, positive pressure; to prevent ingress of dust, sand, or pollutants;
    (b) −ve, negative pressure; to prevent the escape of uncontrolled emissions;
    (c) 0, neutral pressure; where there are no concerns.
2. Stated numbers of air changes per hour (ACH) are lowest acceptable values and indicate the minimum air interchange rates, incorporating both fresh air and recirculated air. Air change rates may need to be calculated to account for:
    (a) removal of excessive heat build-up (equipment and personnel protection);
    (b) the dilution and removal of room air-borne contaminants;
    (c) the provision of combustion and ventilation air supplies for equipment;
    (d) room air leakage where applicable.
3. Noise ratings (NRs) measured at centre room positions, with HVAC system in operation and production/area facilities and personnel at rest. Noise rating curves in accordance with ISO standards specify a noise level for each octave band, and a spectrum noise analyser has to be used to check compliance. However, use of a simple dB(A) noise meter will give a close approximation in accordance with the table below.

Table 4.2 ISO noise rating curve and dB(A) equivalents

Noise criteria	Equivalent reading					
ISO NR	40	45	50	55	70	75
dB(A) equivalent	48	53	58	62	77	82

4.2.2 The mental capacity of humans

It is important to match the design to the mental ability and skills of the operators and maintenance staff. A good example to illustrate this is the design approach adopted for the US Apollo space programme.

- The selected astronauts had to be experienced test pilots and have the ability to assimilate data rapidly and take corrective action in the case of aircraft malfunction. They were also highly educated with science and engineering degrees.
- They were subjected to mental and physical training to ensure they were fit to fly in space.
- They were involved as part of the design team in the design of the spacecraft and its control systems.
- They were involved with all the functional testing and all test phases of the Apollo programme up to the final lunar mission.

A similar approach is taken in the design, construction, start-up and operation of a new process plant. Experienced operations and maintenance staff are involved in the design stage. The appointed operators attend the factory testing of equipment and help in the functional testing during construction. This is followed by maintenance and operations training during start-up on site prior to handover for operation.

The range of attention to ergonomics in design can be illustrated by the following extremes, where attention to safety is matched to the requirements of a particular application in terms of cost and consequence:

- A fighter aircraft requires the maximum integration of man and machine. The pilot has to pass through a highly selective process with intense training to qualify. The cockpit displays and controls must ensure immediate assimilation and action by the pilot. A library of maintenance instructions has to be provided for work to be carried out by highly qualified and trained maintenance staff.
- A motorized lawnmower is designed so that it can be operated by anyone. However, the controls are designed for the average person. No operator training is given. Only a booklet is provided which gives operating and maintenance instructions.

4.2.3 The human control loop

A machine or process is designed to fulfil a function and humans are required to monitor its performance and make adjustments or intervene in the event of malfunction. The process of human control is illustrated in Fig. 4.2.

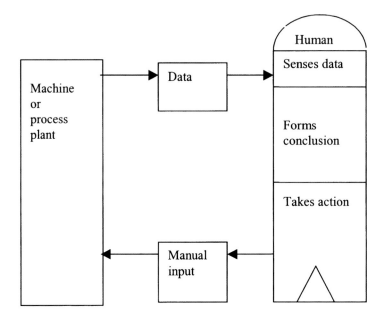

Fig. 4.2 The human control loop

4.2.4   Information: not too much or too little

If the operator is to perform his work efficiently, the data that are provided must be unambiguous and sufficient so as to lead to a clearly defined course of action. Too little or too much information will lead to errors.

In the design of a control system, data are usually required for diverse purposes and they must be grouped and segregated accordingly. The design must ensure that relevant information is displayed. If there is too much, the operator has to spend time deciding what he can disregard and what he must take action on. Mistakes can easily happen.

The sorts of data required from control systems are shown in Table 4.3. With the advent of information technology and computer control systems, more and more information is available to a whole range of people. Screen design becomes important in filtering out each level of requirement. The levels to be provided can be:

- overall plant level with unit general alarms;
- unit level, which provides more detailed alarms;
- controls level for operator intervention;

- operations management level for process changes;
- management level for productivity statistics;
- engineering level for trouble shooting data.

Some levels may require password access for security.

Table 4.3 Data segregation

Type of data	Purpose	Action by
Utilities data	Ensure output is to requirements	Operators:
Processing data		Process adjustments
Fuel/feed input	Malfunction and emergency alarms	Unit shut-down and isolation
Output quantities/quality	Warning alarms	Emergency shut-down
Efficiency data	Defect analysis	Engineers:
Vibration and other condition monitoring data; trends to indicate need for maintenance	Warning of malfunction Maintenance planning	Plan maintenance Testing and defect correction
Reliability data	Production forecasts	Management:
Production statistics	Economic analysis	Logistics
Raw material stocks		Inventory control
Products stocks		Financial control

Even if data are segregated, due to the nature of plant and machines, an impossible array of alarm lights can occur due to cascade effects. For example, the loss of cooling water to a condensing steam turbine will cause loss of vacuum in the main condenser and in the gland steam condensers. Power output will be affected. It may affect the condensate level in the hot well and so affect the boiler feed pumps. It will also cause a rise in lubricating oil temperature. A first-up alarm should enable the operator to realize that the main problem is the loss of cooling water and that all the other alarms are as a result of the loss of cooling water. It may not always be that easy to interpret and too much information can cause the operator to become confused.

Some examples of poor engineering

- An engineering inspection of a process plant revealed that the operators had removed many of the alarm lights. They were fed up with warning lights that they had no control over.
- Another audit discovered that a plant had too many unnecessary trips. This led to lost production without any gain in safe operation. As a result, during the design phase of a new plant, a committee of experienced engineers were charged with the task of eliminating all unnecessary alarms and trips.

Clarity of information

In considering what data are to be provided, the design team must consider all phases of operation and what action is expected from the operator during:

1. start-up;
2. normal shut-down;
3. emergency shut-down;
4. normal operation;
5. part-load operation;
6. each failure mode.

Some typical computer screens for a gas turbine, designed to segregate information and avoid information clutter, are shown as follows.

Gas turbine 1 (Fig. 4.3)

This screen, selected from the main menu, shows No. 1 gas turbine of a three-turbine power station. The station is not running but the screen provides all the information needed by the operator. The array of screen buttons on the left enables the operator to obtain more detailed information as needed. The message strip at the bottom displays any alarm.

Gas turbine 1, proximity vibrations (Fig. 4.4)

Should an alarm message be displayed showing high vibration, clicking on 'Proximity Vib.' on the screen will then display the detailed information as shown. The operator will then see which bearing has the high vibration and if any of the others are affected.

Spinning reserve monitoring (Fig. 4.5)

This screen is also selected from the menu by clicking on the 'Spinning Reserve' button. The station is designed for two gas turbines in operation with one spare. Based on the required load the operator is allowed to arrange the proportion of load for each machine and to decide how many machines should be in operation.

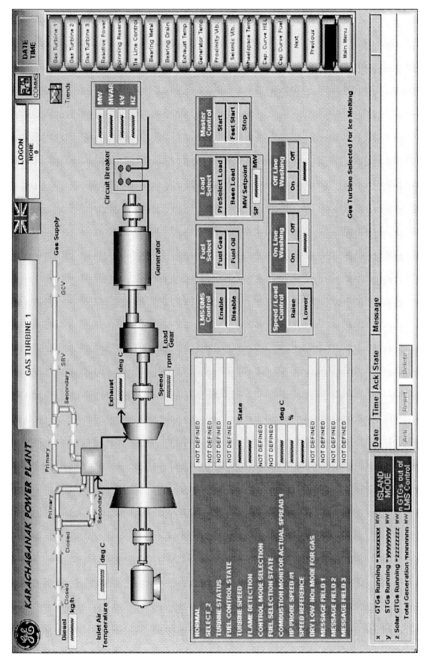

Fig. 4.3 Gas turbine 1

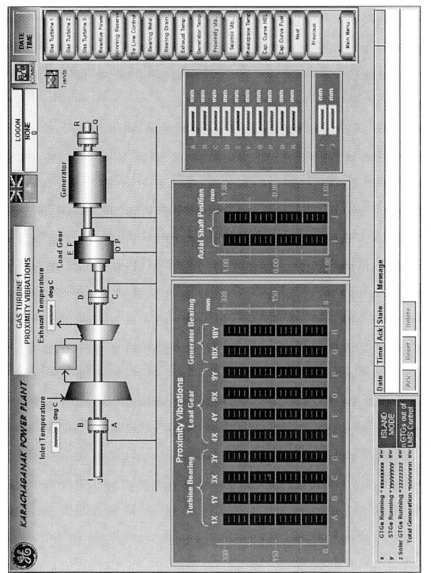

Fig. 4.4 Gas turbine 1, proximity vibrations

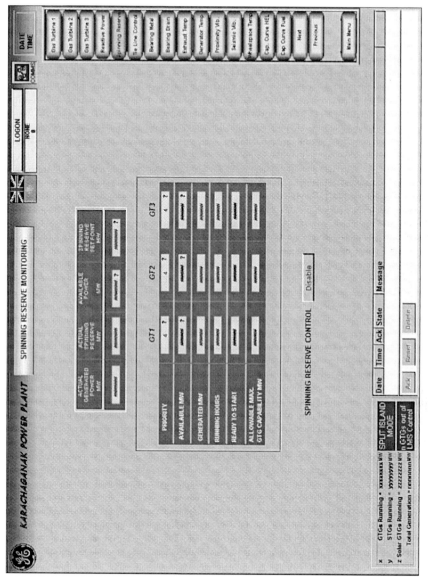

Fig. 4.5 Spinning reserve monitoring

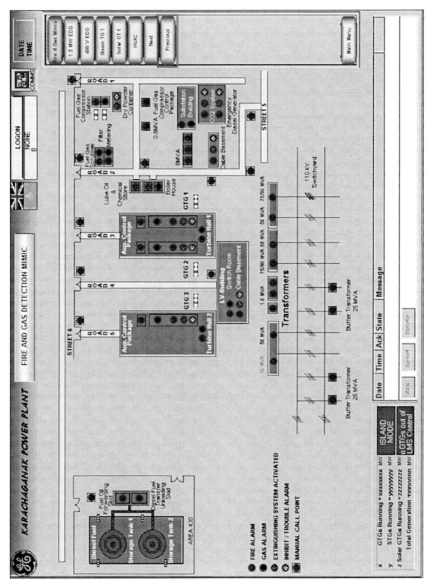

Fig. 4.6 Fire and gas detection mimic

Fire and gas detection mimic (Fig. 4.6)

This is another screen accessible from the main menu. The operator will click for this display to show details of the location of the fire or gas leak in the event of a fire or gas alarm.

4.2.5 Operator controls

The arrangement of controls must be in some logical order, and in some sort of symmetry. This helps to prevent the operator selecting the wrong item to operate when in a moment of panic, or loss of concentration.

- If a push button or valve is located with others that look the same, operator error will occur.
- Clear labelling with good spacing must be provided.
- Geometrical arrangement with different colours may be needed to enhance different functions.
- Where a sequence of operations has to be carried out, a motion study should be carried out during the design phase to ensure that the valves, etc. are located in a smooth logical sequence.

Feedback

To allow adequate control, a feedback signal has to be provided to show the effect of any control action. It is therefore important for the operator to see the effects of his adjustments.

- In pressure venting operations, a pressure gauge should be placed in the line of sight of the operator opening the vent valve.
- In filling operations the operator should be able to see the level gauge.
- The indicating instrument provided for the feedback signal must be located so as not to be confused with any others which may be provided nearby for another purpose.

4.3 Anthropometrics

Human beings come in all shapes and sizes. This is partly dependent on being male or female, and ethnic origin; pigmies and Scandinavians are very different. Designers must take this into account in the location of hand-operated valves, controls and instruments that need to be read.

Car designers attempt to design cars for use by any person, anywhere in the world. As the driving position cannot be fixed to suit everyone, they have had to resort to the provision of adjustable seats and adjustable steering wheel positions.

4.4 Physiology

Humans rely on five senses to orient themselves with their surroundings. These are sight, sound, scent, touch and taste. In riding a motorcycle the rider sees the road ahead. The sound, smell and vibration of the engine and the movement of the machine on the road are all sensed.

In plant operations, usually only the senses of sight and sound are used. These signals have to be processed by the brain, which determines a course of action that results in a physical movement.

The designer needs to consider what time the operator needs to go through this process and whether the operator is able to respond in the time available. If the operator needs to run to a valve that is out of reach and then has to find a ladder, it is going to take some time!

4.5 Psychology

Human beings are not robots. They think and have emotions that affect the way they behave. They respond to the work culture in which they find themselves.

- Training and education are very important ways of conditioning human behaviour.
- Operators working in hazardous conditions, where year in year out nothing goes wrong, will be lulled into a false sense of security.
- A new employee, even if given training, will follow the example of others who have come to belittle the danger.
- When nothing ever goes wrong, a false sense of security can develop, safeguards can get into a state of disrepair and the reasons for them can even be forgotten. For convenience, operators, believing that danger will not arise because it never does, may even render the safeguards inoperative.
- Another fatal flaw is the assumption that the supervisor always knows best, or that someone else is responsible.

These are attitudes of mind that must be overcome by developing a safety culture as a result of education and training. Everybody has to be responsible for safety.

The mental state of humans
Human error can never be totally eliminated and some of the most inexplicable errors are due to an emotional state of mind such as that caused by a death in the family, separation from a partner or a work-related grievance. In these circumstances there could be a 'don't care' attitude that could result in:

- lack of concentration;
- lack of motivation;
- a wilful disregard of instructions.

Emotional factors are difficult to account for and must be detected by supervisor vigilance and sensitivity. Action must then be taken to remove the operator from critical duties until he has a better attitude of mind.

4.6 Quantifying the risk of human error

Human error can never be totally eliminated and there has been much research carried out on how to quantify this risk. Research has established there could be as many as 38 factors to be considered at five different cognitive levels. Various methods for assessing human reliability can be found in reference (1). These methods require considerable knowledge and experience on the part of the user and were originally developed by the nuclear industry. They are also useful in assessing operator risk in process operations.

As an introduction to the subject of quantifying human error a simple method will suffice. This is the method devised by Bello and Columbori, known as TESEO and set out in Table 4.4 (2). TESEO is a method using only five factors and because of this, it is not considered to be accurate. It is only suitable for assessing operator response in a control room type situation and will serve as an introduction to the methods used for quantifying human error. As a demonstration, the method will be applied on the first two of the case histories that follow.

4.7 Case histories

Case 1: Kegworth M1 air disaster, an example of poor information, reference (3)

The first case history involves the wrong interpretation of instruments in the cockpit of an aeroplane. It illustrates a common situation where the operator has to monitor a number of identical items. If one of the items malfunctions the operator needs to know which one and what to do. If it is not clear which one and the operator has not been told what he should do then the chance of error is very high. It is an example of insufficient information and instruction, which illustrates the subject of Section 4.2.4 above.

This is also a good example to demonstrate the application of TESEO and show its validity in such a situation. It will also show that the results can be manipulated to some degree, as the selection of the factors will be a matter of opinion. In this accident the following are the pertinent facts.

1. It involved a twin-engined aircraft.
2. A fault developed on engine No. 1 – excessive vibration.
3. The cockpit instruments – although labelled correctly – were not arranged geometrically correctly. The position of the instruments caused the pilot to believe that it was engine No. 2 that was at fault.
4. The pilot decided to shut down engine No. 2 and the plane crashed with the loss of 47 lives.
5. The pilot had never received training on what to do in the event of excessive engine vibration.
6. It was reported that everyone knew about the poor instrument layout but nothing was done about it.

As an exercise, it is of interest to apply the TESEO assessment of probable human error:

Type of activity is not routine	K_1 is 0.1
Temporary stress factor for non-routine activity (as the pilot was alarmed and not trained, he reacted quickly)	K_2 is 1
Operator qualities – average knowledge and training?	K_3 is 1
Activity anxiety factor – situation of potential emergency	K_4 is 2
Activity ergonomic factor – tolerable interface?	K_5 is 3

Probable human failure can be calculated as

$$P = K_1 \times K_2 \times K_3 \times K_4 \times K_5$$
$$P = 0.1 \times 1 \times 1 \times 2 \times 3$$
$$P = 0.6$$

This means that the probability of error is six times out of ten occasions, a very high risk. Depending on how the K factors are chosen, P could be 1 or even more than 1. This means that, statistically, an error is bound to occur.

In this example, if the engine is vibrating the pilot needs to know how long it can remain in operation or how much further the vibration can increase before he needs to shut down. The pilot also needs positive indication of which engine to shut down and not be left to make his own deduction. If none of these facilities are given to the pilot, he can only make decisions in ignorance of the facts. Too little information and training was given.

Table 4.4 TESEO error probability parameters

Type of activity factor		K_1
Simple routine		0.001
Requiring attention, but routine		0.01
Not routine		0.1
Temporary stress factor, for routine activities		K_2
Time available, in seconds:	2	10
	10	1
	15	0.5
Temporary stress factor, for nonroutine activities		K_2
Time available, in seconds:	3	10
	30	1
	45	0.3
	60	0.1
Operator qualities		K_3
Carefully selected, highly trained		0.5
Average knowledge and training		1
Little knowledge and training		3
Activity anxiety factor		K_4
Situation of grave emergency		3
Situation of potential emergency		2
Normal situation		1
Activity ergonomic factor		K_5
Excellent working conditions and a well designed interface		0.7
Good working conditions and a good interface design		1
Tolerable working conditions and a tolerable interface design		3
Tolerable working conditions and a poor interface design		7
Poor working conditions and a poor interface design		10

Case 2: Herald of Free Enterprise car ferry disaster, the reliability of humans, reference (4) (Fig. 4.7)

The second case history demonstrates the relevance of human psychology, see Section 4.5. It demonstrates that humans are sure to fail. In such a situation, if the consequence of failure is unacceptable then there must be formal supervision and there is a need to enforce a procedure to verify that the person has correctly carried out the duty. The risk has to be managed. Management action is needed to ensure that the importance of the task is reinforced psychologically and that engineering controls are provided to reduce the risk.

Fig. 4.7 Herald of Free Enterprise being salvaged

In the case of this disaster the safety of the car ferry depended on one member of the crew closing the bow doors before leaving harbour. There was no check on this and it was assumed that the door would be closed. This procedure was carried out safely many times with no accident until disaster struck. The pertinent facts were as follows.

1. Due to the need for minimum turnaround time in harbour, the procedure of leaving the quayside before the bow loading door was shut was adopted.

2. One member of the crew had to close the bow door before leaving harbour.
3. Captains were aware of the risk and requested *feedback* to be provided on the bridge to show the door position. This fell on deaf ears and was refused by the board of directors.
4. One day the person concerned was in his cabin and did not know that the ship had left the quayside. No one checked, and the Captain did not know. When the ship left the harbour, water entered through the bow door and the ship capsized with the loss of 188 lives.

There was no one in upper management who made the case that the risk of one person failing was unacceptable because the consequences of failure would be the loss of a ship. It also shows that boards of directors very often do not have members with operational experience who could understand this unless there is someone who can explain it to them.

The question is how much reliance is it possible to place on one man carrying out a routine but critical operation. It can be shown that the TESEO procedure provides a reasonable answer.

Using the TESEO table

$$P = 0.01 \times 0.01 \times 0.5 \times 1 \times 0.7$$
$$P = 0.000\ 035, \text{ i.e. } 35 \text{ per } 1\ 000\ 000, \text{ i.e. } 1 \text{ per } 28\ 571$$

If there were two ships and each did six round trips per day, then between them they will leave harbour 24 times a day. There is likely to be an accident every 1190 days, or 3.34 years.

Note
As there is plenty of time between leaving the quayside and entering the open sea, say 5 min, a K_2 factor of 0.01 has been used.

Case 3: Off-shore crane disasters, beyond human ability
This illustrates the physiological limitations of humans, see Section 4.4. This example shows that it is too much to expect the average human to make split-second judgements in a complex situation and not make mistakes. It takes a highly trained person like a fighter pilot to do this. Otherwise additional instrumentation has to be provided to assist the operator in his decisions. This case history also demonstrates the need for a critical review when a design is to be used for a new application. A study is needed to establish the new operating conditions.

Early North Sea oil platforms were manned with hundreds of people. The regular supply of stores was vital and the platforms were equipped with pedestal cranes, which were used to off-load supply ships. The cranes were constructed according to on-shore designs with not much thought given to the off-shore operating conditions. The correct moment for lifting a load depended on the judgement of the driver. After a number of fatalities, investigation showed that under adverse weather conditions the chance of driver error was very high. The split-second judgement needed led to a high risk of error. The driver had to cope with making too many decisions in the time available. To overcome this, special additional facilities were needed to ensure safety, as explained below.

1. Cabin designs needed to cope with off-shore storm conditions with regard to driver comfort and visibility.
2. In heavy sea conditions, the supply vessel can heave up and down some 5 m. If the load is lifted when the ship is falling the force needed to hoist is equal to the static load plus the dynamic force to overcome the downward motion of the load. This could exceed the allowable load on the crane. The driver also has to follow the position of the ship, which is in constant motion; this affects the load that can be safely lifted because this in turn affects the reach of the jib.
3. To minimize the lifting force the driver must hoist when the supply ship is rising. In poor visibility it is easy for the driver to misjudge this.
4. Besides allowing for dynamic forces, the crane driver also has to note how far out the crane jib has to reach. The further the jib has to reach the lighter the load that can be lifted. In a standard land crane these limits are given as tables displayed in the cabin, which the driver has to consult. With heavy seas there is no time for the driver to work it out – he has to make a guess.

To reduce the risk, drivers' cabs are now supplied with instrumentation to show wind speed, which helps the driver to assess sea state, which in turn affects the likely motion of the supply ship. Manual input of sea state into the control instrumentation by the driver, with automatic feedback from deployment of the jib, then provides an automatic display of allowable lift. The driver is also provided with instrumentation to show the supply ship's up and down motion, to allow the driver to decide when to hoist without any need for guesswork. In addition to all this extra instrumentation, the crane can also be provided with an excess load safety system.

This is an example of engineering controls to make up for human limitations. With the additional instrumentation no further accidents have occurred. In fact, in one case, where the hook got caught up on the supply ship, the safety device prevented the crane from being toppled.

Case 4: Three Mile Island power station, a nuclear disaster caused by poor training and information overload, reference (5)

This example illustrates the points made in Sections 4.2.3 and 4.5 – the requirement for operators to make decisions based on the information provided and the need to provide training as to the response expected.

In this example the information provided was massive and the operators were required to filter the information and then to make the correct decision. It was too much for them and so they made the wrong deductions.

Background information

A pressurized water-cooled reactor produces heat, which is carried away by a pressurized water circuit. The pressurized hot water is used as a heating medium, circulating through a heat exchanger that produces steam to drive steam turbines. In the pressurized water circuit there is a steam drum that has a water level and a steam space. In the event of a turbine shut-down, no steam is used and so the pressurized water will overheat.

The control system recognizes this and the reactor is shut down. However, due to the reactor radioactive decay heat output, it takes time to cool down. This excessive heat causes the pressurized water to boil and increase in pressure. A control valve then opens to release the steam, which also causes the water level to drop as the steam is boiled off. Cold make-up water is added which, together with the release of steam, drops the pressure and so causes the control valve to close. This continues until the reactor is cold.

The events that led to disaster

1. On the day in question the steam turbine tripped due to a problem.
2. The control system initiated the normal reactor shut-down process.
3. The pressurized water circuit overheated, and as to be expected the control valve opened to allow the steam to boil off. However, when it reached the point where the valve was required to close, there was a malfunction and it did not close as required.
4. The instrumentation showed correctly that a close command had been given but no *feedback* signal was provided to tell the operators that the valve was still open.
5. Due to falling pressure and loss of water due to leakage through the open valve, the cooling water reached saturation temperature. This of course caused the water level to rise in the steam drum because boiling water takes up more volume.

6. The safety systems caused the make-up water pump to start up.

This had the following results for the operators.

*   On the initiation of this incident the operators were overwhelmed with some 2000 alarm warnings.
*   They had some tens of danger alarms.
*   When the make-up water pump correctly started up, they were confused because the steam drum indicated high water level and they thought that the leak-off control valve was closed. They therefore concluded incorrectly that the emergency water pump started up in error and they shut it down.

During normal operation it was important not to overfill with water and the danger of overfilling with water dominated their minds. Other signals giving the water pressure and temperature were overlooked because their training had not prepared them for this particular type of failure. They did not realize that saturation had been reached nor did they know what it meant.

The operators suffered from information overload, wrong information, and a lack of understanding as to what happens when the cooling water boils. The control system was correctly designed to safeguard the reactor but was incorrectly overridden by the operators.

Cascade failure, subsequent meltdown of the reactor core, and the leakage of radiation into the environment occurred because the operators intervened and shut down the make-up water pump.

Note

Modern nuclear plants have other ways of controlling plant temperature, which avoids this type of disaster: an example of safety integration.

This example demonstrates the importance of training and the need to ensure that all situations are considered. In an emergency, abnormal situations will occur which may need abnormal actions. Operators need to be trained and exercised to deal with them. It requires risk management.

Case 5: Chernobyl nuclear power station, disaster due to complacency

This illustrates the effect of human psychology, see Section 4.5. People who live in situations with risk become complacent and believe that danger will never arise. This is an example, which occurs all too often, where operators wilfully remove or bypass safety measures and disregard the danger.

This was a USSR-designed water-cooled nuclear reactor, different from those elsewhere in the world. It was known to be unstable and likely to meltdown at below 20 per cent output. The reactor was installed with safety systems to prevent operation below 20 per cent.

The events leading to the disaster

1. In the event of an emergency shut-down of the power station, emergency diesel generators are needed to ensure essential supplies.
2. The emergency diesel generators needed 2 min to reach full power from receiving the start signal.
3. The management wanted to know what power would be available from the reactor at low outputs.
4. They decided to run some tests at low outputs and to disable the automatic shut-down systems to enable them to do so.
5. In their tests they deliberately operated below 20 per cent output.

In removing the automatic safety systems, it would seem that no thought was given to the possible danger of meltdown. In the event, the reactor did become unstable, and, due to other inherent design weaknesses, manual intervention by the operators was too slow to prevent the disaster.

This illustrates very well that safety systems must never be disabled. If proposed it must first be fully considered by experts and approved by government safety authorities at the highest level.

Case 6: Crash landing, disaster due to rigid hierarchy

Another example of human psychology, see Section 4.5. This shows how initiative can be stifled by indoctrination. It also demonstrates the need for management to enforce procedures without exception.

A young pilot was under training with a senior instructor. The senior was an autocratic taciturn sort of character. When coming in to land the young pilot carried out the duty of reading out the landing procedure checklist. The senior pilot made one or two grunting noises and not much else and the plane made a crash landing. It turned out the senior pilot had died at the controls. The young pilot didn't dare to enquire when things were going wrong due to the belief that the senior could not be questioned.

Case 7: A medical tragedy, disaster due to multiple human errors

A further example of human psychology, see Section 4.5. This shows that when more than one person is involved, everyone does nothing, thinking that someone else has the responsibility. It demonstrates the need for risk management and the development of a safety culture where everything is assumed to be wrong until it has been checked to be right.

The events that led to disaster

A newly appointed Consultant Registrar was doing the hospital rounds with his team. Examining a patient who had an infection, the Consultant Registrar instructed that an antibiotic should be prescribed. The Medical Officer in attendance suggested a penicillin type to which the Consultant Registrar agreed. The Medical Officer wrote the prescription on the medication card. The ward staff subsequently administered the medication and the patient died.

The Consultant Registrar was charged with gross criminal negligence. The medication card had a warning notice in red that the patient was allergic to penicillin. Everyone ignored or did not read the notice. The Consultant Registrar relied on others to read the notice and they all failed.

This example illustrates the fact that relying on more people to check does not reduce the risk; in fact on occasion it can increase risk because everyone assumes that others have done the work. The principle of redundancy to reduce risk, as shown in Chapter 8, is only valid with machines, not with people.

4.8 Summary

In the wake of any disaster, usually the first reaction is to blame the operators. People make mistakes. As has been shown, poor design or bad management could have been more to blame.

Some operators do deliberately take the wrong action due to perverse reasons. Mostly operators make what they thought was the best decision at the time. They were guided to this decision by the information they were given and the training and education that they had received.

Poor design, training and education is not the responsibility of the operators. It is the responsibility of engineering and management.

Human error needs to be reduced by engineering controls and the establishment of good working practices. The possibility of human error has to be recognized and the risk managed.

4.9 References

(1) **Humphries, P.** (Ed) (1988) *Human reliability assessors guide*, RTS88/89Q, NCSR, UKAEA, Warrington.
(2) Attributed to: **Bello, G.C.** and **Columbori, V.** (1980) *Reliability Engineering*, 1(1), 3. Taken from: **Kletz, T.A.** (1991) *An engineer's view of human error*, Second edition, IChemEng, ISBN 0 85295 265 1.

(3) Trimble, E.J. (1990) *Report on the accident to Boeing 737-400G-OBME near Kegworth, Leicestershire on 8th Jan 1989,* HMSO, London, ISBN 0115509860.

(4) Sheen, J. (1990) *'M.V. Free Enterprise' Report of court No. 8073,* HMSO, London, ISBN 0115508287.

(5) Kemeny, J.G. Chairman (1979) Report of the President's Commission, *The accident at Three Mile Island,* Pergamon Press, ISBN 0 08 025946 4.

Chapter 5

Safety Integration

5.1 Introduction

Knowing or identifying hazards has been dealt with in the previous two chapters. This chapter is about how to deal with the resulting risk. There is a hierarchy of preference to hazard risk control, which is:

1. alter the design to avoid the hazard;
2. provide facilities to reduce the risk from the hazard by design;
3. provide procedures to protect exposed persons;
4. provide means for personnel protection.

Safety integration is the provision in a design to provide risk control of hazards. Ideally hazards should be eliminated by design in accordance with the hierarchy of preference given above. Examples of the application of the different levels of the hierarchy will be given for various hazards.

In many situations the hazard is an inherent part of a process; for example, in an oil refinery the hazard of fire and explosion cannot be avoided. However, the risk of fire and explosion will be specific to particular process areas. Risk control has to be considered at the start of design and the layout of the plant is critical in ensuring avoidance of risk to people. Avoidance of risk to people is achieved by ensuring that facilities such as office buildings, stores and workshops are located away from high-risk process areas. With the advent of computerized controls and CCTV, control rooms can also be remotely located.

Storage tanks with flammable fluids will need to be as far away as possible from areas with risk of fire. Where control rooms have to be close to hazards, designing them to be fire and blast proof with suitable means of escape provides protection for operators.

General principles for the application of risk control by design are given below. They will serve as an introduction to the understanding of established codes and standards. Most will also be covered by regulations that must be studied to ensure compliance.

5.2 Hazardous area classification

There are many types of plant and equipment that process or use flammable gases. To prevent fire and explosion, it is necessary to prevent its ignition in the event of any gas leak. In the design stage, it is usual to identify the areas where gas can leak as a hazardous area. Apart from ensuring that any naked flames are not in these areas, it will also be necessary to ensure that no electrical arcing can take place.

The basic principles for establishing the risk of ignition are:

- likelihood of release   Zone or Class classification
- type of flammable material   Group
- temperature of ignition   T classification

The two major internationally recognized codes of practice are API RP 500 issued by the American Petroleum Institute and the IP code Part 15 issued by the Institute of Petroleum.

After 30th June 2003, the EEC ATEX 99/92 directive will be applicable in Europe. The definitions of IP code Part 15 would appear to be adopted and extended to include other industries that are subject to explosive dust clouds.

5.2.1   Likelihood of release

Zone or Class classification based on the likelihood of release is given in Table 5.1. It can be seen that the two codes are similar. The descriptions given are abbreviated and the API RP 500, Section 4.2.1b, NEC (US National Electrical Code) 500-5(b) and the IP code Part 15 should be consulted for a more complete definition. The interpretation of how far a hazardous area extends – upwards, downwards and sidewards – from any source of leak is governed by rules that differ between the two codes, and these must be studied for any given application.

The rules given in ATEX will need to be checked for any differences that may have been introduced to ensure compliance.

5.2.2   Type of flammable material

These are grouped on the basis of their hazardous characteristics, which depend on how easily they can be ignited. The US NEC system arranges flammable materials in groups ranging from Group A, the most easily ignited, to Group E, the least. The European International Electrotechnical Commission (IEC) system uses two groups. Group I is below ground, i.e. the mining industry, and Group II is for above ground. Group II in turn is subdivided into subgroups based on the energy needed for ignition: II A, high energy; II B, less energy; II C, little energy (50 mJ).

Table 5.1 API code and IP code classifications compared

API RP 500		IP code Part 15	
Class	Definition of location	Class	Definition of area
Class 1, Division 1	Ignitable concentrations of flammable gas are expected to exist or where faulty equipment might release gas and cause failure of electrical equipment	Zone 0	Where a flammable atmosphere is continuously present, or present for long periods
Class 1, Division 2	Ignitable concentrations of flammable gas are present, but are confined, or prevented from accumulation by adequate mechanical ventilation, or are adjacent to a Division 1 area from which gas could occasionally be communicated	Zone 1	Where a flammable atmosphere is likely to occur in normal operation
		Zone 2	Where a flammable atmosphere is not likely to occur in normal operation and, if it occurs, will only exist for a short period
	ATEX directive extension		
		Zone 20	Where a flammable atmosphere in the form of a combustible dust cloud is continuously present, or present for long periods
		Zone 21	Where a flammable atmosphere in the form of a combustible dust cloud is likely to occur in normal operation
		Zone 22	Where a flammable atmosphere in the form of a combustible dust cloud is not likely to occur in normal operation and, if it occurs, will only exist for a short period

5.2.3   Temperature of ignition

Liquids that vaporize when heated can ignite. It should be noted that the temperature at which they turn to vapour, the flash point, is not the ignition temperature, which is much higher. Gases react differently and can explode when mixed with air and ignited.

To prevent auto-ignition, electrical equipment is subject to standards that regulate the maximum allowable working surface temperature. These are called T classes, which range from T1 to T6. The T class applicable is shown in Table 5.2 for the materials classified.

Table 5.2 Gas groups and classes, European and US

Typical gas	European IEC 79 BS 5345 (UK)	USA NEC 70 Class group	Ignition temperature	T class (maximum allowed surface temperature)
Aluminium and magnesium/ alloy metal dust		E		
Methane	I	D	650–749 °C	T1 (450 °C)
LPG, propane, natural gas, petrol (gasoline)	II A	D	354–366 °C	T2 (300 °C)
Ethylene, ethyl–ether vapour, cyclopropane	II B	C	543 °C	Ditto
Hydrogen	II C	B	524 °C	Ditto
Acetylene	II C	A	404–440 °C	Ditto

5.2.4   Equipment for hazardous areas

Equipment must be in compliance with applicable design standards and also certified by a recognized authority (a notified body, for example BASEEFA/EECS or TUV) in accordance with the ATEX directives. In the USA, equipment must comply with NEC/NEMA regulations and be certified by a recognized approval body such as Underwriters Laboratory (UL) or Factory Mutual (FM).

In Europe these requirements cover electrical equipment, internal combustion engines and protective systems. This is at present more extensive than in the USA.

Again it should be noted that the work place ATEX directive 99/92 comes into force in June 2003; retrospective certification of existing equipment in use in hazardous areas, as applicable, will be needed by June 2006. The directive will also extend the requirement for the classification of electrical equipment to mechanical equipment. Both mechanical and electrical equipment will need to be certified as follows:

Hazardous zones:	Equipment to be certified:
Zone 0 and Zone 20	Category 1 and M1
Zone 1 and Zone 21	Category 1 or 2 and M1 or M2
Zone 2 and Zone 22	Category 1 or 2 or 3

5.3 Fire prevention

Design features to reduce the risk of fire may be subdivided into groups as explained below.

5.3.1 Segregation

This is the principle that sources of possible fire hazards should be separated from combustibles. Firebreaks should be formed and so prevent propagation in the event of a fire. They should also be separated from people and locations of high value. A spacing that has been typically used for refineries is given in Table 5.3. The actual spacing adopted will also be influenced by the installation of fixed fire protection equipment balanced by the expected risk of a fire.

In Table 5.3 no figures have been included for storage tanks because the rules differ depending on whether they are of 8000 m^3 capacity, or below or above this. Large tanks have different rules depending on their construction. For example, large floating roof tanks up to 45 m diameter should be 10 m apart and those above this size should be 15 m apart. Depending on the risk of ignition and if space is limited, fixed fire protection may be necessary. The HSE issues guides on this. The IP model Code of Safe Practices, Part 19, gives guidance for large tanks.

The same principles apply to the design of buildings, warehouses and stores; consideration will need to be given to the identification of hazards. Can the hazard be moved elsewhere with less risk to people? If not then design features will be needed to reduce the risk from the hazard. The principles of segregation, detection and control will then need to be applied. BS 5588 Fire Precautions in Buildings and Structures should be consulted for separation requirements.

Table 5.3 Typical industrial spacing (m)

	Item	A	B	C	D	E	F	G
A	Office, laboratory buildings, etc.	3						
B	Process units	50	25					
C	Stores with flammable materials	25	25	15				
D	Air intake and other sources of ignition	3	25	25	1			
E	Liquefied gas storage	50	25	25	25			
F	Crude oil storage	50	25	25	25			
G	Flammable liquid storage tanks	50	25	25	25			
H	Site boundary fence	15	25	15	3	60	60	60

5.3.2 Detection

Fire detectors

This is a design measure to reduce the risk from fire; early detection and alarms allow people to escape. The linking of detection signals to the automatic initiation of fixed fire fighting systems will prevent escalation.

In the use of detection systems the issue of reliability is paramount. Initiation of fire fighting systems if there is not a fire is just as bad as if the detection system fails to operate if there is a fire.

Detectors sense the effects of a fire according to smoke, heat and radiation. They must be selected and positioned according to the type of fire and flammable material at risk. The principle types and their features are given in Table 5.4. As can be seen, there are many types available and some judgement is needed in their selection. Each has its advantages and disadvantages, and a mix and match may be needed, based on the type of fire expected and the type of flammable material involved. The use of diverse methods of detection will also improve the reliability of detection. EN54 prescribes fire tests for testing the sensitivity of detectors to different types of fires and classifies them with regard to their sensitivity.

Table 5.4 Fire detectors and their use

Detector type	Features
Smoke detector	Responds to both visible and invisible products of combustion. Typically used for offices, and commercial and residential buildings. Oil vapour can give false alarms.
Carbon monoxide (CO) detector	Responds to CO which may be generated before there is smoke. Good for areas for accommodation and large spaces such as cargo holds, theatres. Immune to typical smoke detector-type false alarms.
Fixed-temperature detectors	These have a preset temperature, but are slow in response. They are fitted to sprinklers. Thermocouples are another example.
Rate-of-temperature-rise detector	They respond to a rise in temperature, with a fixed maximum temperature setting. They are faster than fixed-temperature detectors. In areas such as engine rooms, a sudden rise in ambient temperature can cause spurious responses.
Rate-compensated heat detectors	These have a fixed temperature setting which drops to a lower setting if there is a rapid temperature rise. They are not susceptible to a rapid change in ambient temperature.
High-performance optical detector	This combines the rate-of-rise detector with an optical smoke detector. Normally the smoke sensor sensitivity is low to avoid false alarms. A rapid rise in temperature signal is then used to increase its sensitivity. An alarm is only given if smoke is detected.
Ultraviolet flame detectors	They are immune from solar radiation and only respond to ultraviolet light given off by a fire. They respond to ultraviolet light from arc welding and sometimes from quartz halogen light. They are blinded by hydrocarbon deposits and smoke on the lens.
Infrared flame detectors	They respond to infrared rays given off by burning carbon and use filters to avoid the effects of the sun or hot surfaces. They can react to reflected flickering sunlight, e.g. off water, and can be blinded by icing.
Triple wavelength infrared flame detectors	One unit senses CO_2 emission and the other two sense the background infrared level. Signal processing is used to process the three signals and to determine if a true alarm exists.
Combined ultraviolet and infrared detector	This is two units in one to combine the advantages of both. The only disadvantage is a higher cost.

CCTV smoke and alarm detection system

This uses special software to compare one TV frame with the next so that any frame can be evaluated. The algorithm used is able to identify large clouds of thin smoke as well as small areas of thick smoke. Based on detecting the change of light attenuation, the evaluation is carried out every second and provides an automatic alarm within seconds. The system can detect leaks of steam or oil vapour as well as smoke. The operator looking at the CCTV monitor can verify the cause of alarm.

Gas detectors

Gas detectors are available that will detect flammable gases. They are usually set at some lower explosion limit (LEL): one at 25 per cent LEL for alarm and one at 50 per cent LEL for trip. With time they become contaminated and are unreliable. For this reason defect monitoring is provided and it is usual to install three for a two-out-of-three voting system. Optical infrared gas detectors are also available which are not susceptible to poisoning and so are more reliable. Infrared beam detection may need to be used in outdoor environments where gas clouds are affected by wind.

Toxic gas detectors are also available; the setting for these will depend on the toxicity of the gas and the threshold limit values and short-term exposure limits as usually given on the associated safety data sheet.

Oil mist detectors

These are required to be fitted to the crankcases of large marine engines to provide an alarm and avoid any possibility of a crankcase explosion.

Multi-detector systems

The availability and use of programmable computers to receive and process multiple signals have enabled the use of fire detection algorithms. By using data that characterize the development of different types of fires, it is possible to eliminate false alarms and provide a rapid response to a real fire.

5.3.3 Suppression

Should a fire be detected, preinstalled fire fighting systems can be in place to put out the fire, see Table 5.5. This is a design provision for protection from the fire hazard. It allows time for the arrival of the fire fighters and prevents any propagation. The provision of these services must be considered early in the design phase so that their location and routing can be considered during the layout of the plant. Where the use of water or foam is planned, then the provision of adequate drains to carry away the water in the event of a fire will be needed.

Table 5.5 Types of fixed fire protection and their application

Type of protection	Description
Water spray	An array of nozzles supplied with water from a grid or network of pipes. The mains water supply can also supply a number of grids with zone valves to select which are to be activated. When operated all nozzles discharge simultaneously.
Automatic sprinkler system	As above except that each nozzle operates individually, activated by fixed-temperature detectors.
Foam system	This discharges foam (instead of water) through a sprinkler system. A fire fighting foam concentrate is proportioned into the water supply to produce the foam.
CO_2 system (causes lack of oxygen, note safety hazard)	An array of nozzles supplied with CO_2 from a grid or network of pipes. The CO_2 is released from a battery of storage bottles which then supplies the network via a mains supply pipe. Just as in a water spray system, a central supply can be used to supply a number of zones.

Hazard	Type of system used
Ordinary combustibles, wood, paper, etc.	Automatic sprinkler system.
Rack storage	Automatic sprinkler system. Specially designed to suit the storage racks.
Plastics	Automatic sprinkler system. Beware of toxic fumes!
Flammable liquids	Water spray system. Low-flash-point liquids will need a foam system.
Flammable gases	Water spray or sprinkler system. To block radiation and dissipate heat until gas flow can be isolated.
Electrical	Use CO_2 if warranted. Beware of electric shock if water or foam is used! Use water spray for oil-filled transformers.
Combustible construction	Where plastics, etc. are used, use water spray system.

5.3.4 Hazards from CO_2

The use of CO_2 to put out a fire is in itself hazardous. It works by reducing the oxygen content in a room. When fire is detected, the HVAC must be automatically shut down, all the ventilation dampers closed and the CO_2 discharged. Design provision must be made to avoid the hazard.

If a person is trapped in the room, death can occur. As a safeguard, facilities must be made available to turn off the automatic discharge of CO_2 while people are present. The system is then placed under manual control. In the event of a fire, the people, on leaving the room, activate the system manually. A system of indicator lights, together with the lock off and manual activation facilities, should be located at the entrance to the room.

5.3.5   Avoiding CO_2 hazards – water mist fire suppression

The hazards of CO_2 and the problems of using water deluge systems at sea can be avoided by design. (The use of firewater saved a ship from fire, but the water caused instability and it capsized.) This has resulted in an alternative method being developed. This system uses very fine water droplets on the basis that the heat gain will cause them to boil and evaporate. The effectiveness of the system depends on the droplet size being between 50 and 120 μm. The system is SOLAS-approved for local application, and has the following advantages:

1.   provides a cooling effect;
2.   inerting effect at the fire due to the drops flashing to steam and so displacing the O_2;
3.   radiation blocking due to the water mist;
4.   causes minimum damage to equipment.

The system is suitable for electric, gas and oil fires and can be used instead of CO_2, powder or foam. Systems are available for computer room fires where the main damage is caused by smoke. Due to the need for a very small droplet size, nozzles with integral filters are provided to prevent clogging, and strict cleanliness is needed.

5.3.6   Containment

Fires when they occur must be contained to prevent their spread and so minimize risk. Design provisions for fire resistant walls, fire retardant doors or other methods of containment will reduce risk.

- In test cells, for example, the building construction can be done on the basis that any fire is prevented from spreading to the adjacent cell.
- The fuel tanks for an engine room can be located in a separate room.
- Fuel tanks should be surrounded by a bund high enough to contain the contents in case of rupture and to prevent any flow of burning fuel in the event of a fire.

5.3.7   Means of escape

These are provisions to protect exposed persons. Buildings can be located at a safe distance from plant but they too have a risk of fire, albeit a small one. Operators are needed to patrol plant areas and maintenance crews may also be working in plant areas. They will be at risk. All design layouts should be checked to ensure that people couldn't be trapped without a means of escape. Normal situations and emergency situations must be considered and the means of escape verified to check that they cannot become blocked.

It is always necessary to have two routes available, and the distance to any one of them should not be excessive. Large rooms must have two exits. The escape doors must open in the direction of travel and the route must always lead to a safe location at ground level outside the building or structure. In special situations, routing to a place of refuge is an acceptable alternative, so long as there is a means of rescue from that location.

All escape routes and exits must be clearly marked, complete with emergency lighting that can still operate in the event of the loss of power.

5.3.8   Emergency shut-down (ESD)

In the event of any fire, a process plant will need to shut down safely. In doing so, the following objectives must be met:

1. the shut-down must be in an ordered and safe sequence;
2. any feed streams to the seat of any fire must be predetermined and be automatically isolated;
3. any failure of equipment due to the fire must not result in the release of anything harmful to the environment;
4. any pressure vessels must be isolated and vented down to avoid an explosion due to being heated up;
5. confirmation that all initiated actions have been completed.

In an emergency, it will be impossible to expect the operator to remember all the different actions needed to accomplish the stated objectives. An ESD procedure must be determined in advance and programmed into a computer control, which is activated by a special ESD push button to shut down the plant. These are provisions in design to avoid possible operator error. They ensure that the measures to minimize the risk of fire and explosion are reliably carried out.

5.3.9   Security

Although all the design safeguards have been provided, the final design check that has to be made is to ensure that the safety facilities cannot be destroyed in the event of a disaster. This is to ensure that the facilities provided to reduce risk can be relied upon.

- Firewater pumps and firewater storage facilities may need to be duplicated and segregated to ensure their availability. Both diesel and electric motor drivers will need to be used for diversity and to avoid common failure due to loss of electric power.
- Firewater mains may need to have alternative routes and be buried to ensure security of supplies.
- Electric supplies to emergency services must be duplicated from two different sources and by two different routes.
- Control and communication cables will also need duplication and segregation to ensure their survival.
- Control rooms may need to be blast proof to ensure that they remain in operation.

5.4 Design to ensure safety

Besides the hazard of fire, there are many other common hazards to be considered and some examples are given below.

5.4.1 Explosions

There may still be the hazard of an explosion, even after all provisions have been made to reduce risk. The residual risk can be controlled by the use of blast walls, or blow-out panels if the hazard is in a building. This channels the direction of the blast in a safe direction. At one time crankcase explosions occurred in large marine diesel engines and ships caught on fire and even sank as a result. Investigations revealed that the overheating of bearings caused the explosions. If the crankcase oil was also contaminated with fuel the hot bearing could vaporize an explosive mixture and ignite it. Design provisions removed this hazard. Crankcases were fitted with blow-out doors and flame arresters. This controlled the explosion and prevented any fire. In modern engines, besides blow-out doors, the bearing temperatures are continuously monitored and so the hazard can be avoided.

5.4.2 Falling

Falling causes some 56 per cent of industrial injuries. The hazard of falling can be avoided if, during design, some thought is given to the location of equipment. In HVAC installations, for example, it is quite common practice not to consider the location of instruments and leave their location to chance during installation. On one project, checking by the client revealed that the locations were totally unacceptable because of poor access and the need for elevated maintenance. When this relies on the use of ladders and temporary platforms, there will be a high risk of falling. Any fall from above 2 m can result in major injury. Even falls less than 2 m can result in a lost-time injury. The first priority is to install equipment as low as possible, to be less than 2 m high. If it has to be located higher, the risk can be avoided by facilities to remove and lower the equipment for maintenance or to provide fixed ladders and platforms. A risk assessment will need to be made with regard to frequency of access balanced against the cost of the facilities. Other provisions could then be considered.

Many falls could be at ground level due to slipping on an oily surface. API standards for machinery and oil systems recognize this and require all base plates to be of the 'drain gutter type with one or more drain connections of at least 38 mm in size.' Furthermore the API requires that 'non-slip decking shall be provided... covering all walk and work areas.' This is an example of avoiding the risk by design.

5.4.3 Equipment lifting

Lifting accidents account for 5 per cent of all industrial accidents. Practically all maintenance operations require some form of lifting. Provision of proper lifting facilities can reduce the risk of improperly secured loads falling. Here again the API leads the way, in requiring lifting lugs to be provided for all casings that need lifting for main-tenance. Besides increasing safety, lifting provisions will improve plant reliability, as they will reduce the mean time to repair (MTTR).

Besides the need for lifting attachments on all items that need lifting, there is also a need for facilities to lift. The hazard from the use of inadequate lifting arrangements can be avoided by making the proper provisions available. For small loads simple provisions, such as locations for hoist attachment, should be provided. For larger loads, beams for movable hoists will be needed. For major equipment, travelling cranes will have to be provided. Lifting capacity must match all loads to be lifted and all facilities must be adequately labelled as to their capacity.

In the planning for travelling cranes, a survey of the lifts and movements needed should be undertaken to identify any hazard that could arise. There could be danger of collision with other equipment and a system of limit switches on the crane rails may be needed to avoid any risk of traversing into an obstruction with the load at an incorrect elevation. Consideration of the consequences of dropping a load and its impact on safety and collateral damage must be carried out. Design provision for its correct location to minimize risk and avoid hazards can then be provided in accordance with the principles of risk control.

5.4.4   Motion of machinery

It is well recognized by safety regulations that moving parts are a hazard and that guards are needed to prevent inadvertent contact. A hazard that may not be so well recognized is the inadvertent movement of machines when shut down for maintenance. It is important that machines are prevented from moving while people are working on them. Large machines are big enough to allow people to work inside unseen. The hazard that the machine could move, with fatal consequences, is well documented. This hazard can be avoided by design, with facilities to lock the motion works and prevent movement. Large machines need barring gear to enable the machine to be rotated manually. Often this can also be used to lock the machine in a set position. Starting systems should also be isolated. This is automatic if an interlock is provided that will prevent the starting system from being activated when the barring gear is engaged.

5.4.5   Entry into enclosures

Entry into tanks and vessels and other enclosures is required for inspection and maintenance. This is dangerous if the atmosphere is hazardous.

This hazard can be avoided if purging and testing facilities are provided, either in the form of a permanent installation or facilities for the connection of temporary facilities. In confined spaces there could be the possibility of entrapment or engulfment. Design to provide installed rescue equipment and facilities to prevent unauthorized entry will reduce the risk of fatalities.

5.4.6   Transfer of hazardous materials

The hazard of spills and splashes can be avoided by using mechanical transfer by pipes from bulk storage, designed to avoid human contact. If manual handling cannot be avoided, the use of transfer pumps will reduce the risk of contact. In spite of protective gear people can get splashed. Safety showers and eye baths are required to provide first aid if needed. Provision of containment areas for transfer operations, with disposal facilities, will help to contain and minimize the hazard from any spill.

Oil tanker transfer operations are hazardous. Moving away while connected, or being disconnected before closing isolation valves, will result in spillage and the risk of fire. The risk is avoided by the use of breakaway, auto-closing couplings and automatic ESD.

5.4.7   Diesel engine fires

There have been many engine room fires caused by fractured fuel pipes. Any leak will result in a high-pressure spray that can vaporize and ignite should it impinge on to a hot surface. Heavy low-grade fuel is often heated to 2.5 times the enclosed flash point and, on leaking under pressure, will produce a large volume of flammable vapour. The best way to stop a fire is to prevent any fuel leak. Sheathed metal fuel pipes are now fitted on marine engines. The outer sheath retains any leak from the pressurized inner pipe and the leaked fuel is drained into a reservoir, which is fitted with a liquid level alarm. This is a good example of safety integration where the risk has been avoided by a design change.

5.5 Designing for system reliability

Some of the features required for safety also improve reliability. In this section, some features that are used to improve reliability are shown. Sometimes they can have an adverse effect on safety as shown in the following example.

5.5.1 Dual pressure relief valves

As previously discussed, the reliability of pressure relief valves (PRVs) depends on the time interval between testing. In some critical situations, on continuous process operations, dual pressure relief valves are installed. The advantage is that relief valves can be removed for testing without stopping operations. The disadvantage is that this can in itself pose a hazard to safety. Examination of the procedure needed to change a PRV shows how mistakes can be made.

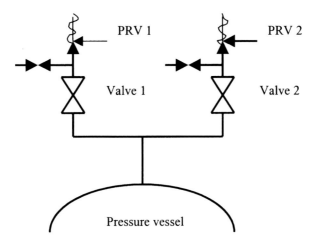

Fig. 5.1 Dual pressure relief valve installation

Possible errors (Fig. 5.1)

1. Close valve 1 before opening valve 2, the vessel will be without a PRV while valve 2 is closed.
2. Close valve 1 and forget to open valve 2, the vessel has no PRV.
3. Open valve 2 and forget to close valve 1, a possible fatal injury in attempting to remove PRV 1.

Maintenance procedure

Normal operation

- PRV 1 in operation;
- valve 1 is normally open, with vent valve shut;
- valve 2 is normally shut, with vent valve open;
- PRV 2 has been removed and tested and reinstalled.

Change-over operation

- PRV 1 is in operation;
- valve 1 is open, with vent valve shut;
- open valve 2 and close its vent valve;
- PRV 2 is in operation;
- close valve 1 and open its vent valve and remove PRV 1 for testing.

Comment

It can be seen that in improving reliability, hazards to safety have been introduced. Engineering changes are needed. One provision is by the application of a mechanical interlock system that depends on a series of trapped keys. The first key is held in the safety office. When a permit is issued to change over, the key is handed over to the technician. This key enables valve 2 vent valve to be closed. When valve 2 vent valve is closed, the first key is trapped, but a second key is released which allows valve 2 to be opened and so on until PRV 1 can be removed safely. The final key that is released is given to the safety office for reissue at some future time.

A design provision to avoid the hazard is to use a three-way through-flow ball valve, which allows switching over without ever blocking off a PRV.

Modern safety selector valves of special design are now available which incorporate all the required features in one integrated mechanism. This then ensures complete safety in the removal of PRVs.

Another very common requirement is the isolation of one of a number of similar pressure vessels. There will be isolation valves at the inlet and at the outlet together with a vent valve. It is quite easy to open and close the wrong valves, especially if the valves are not positioned in such a way that it is obvious for which vessel the valve is intended. Opening and closing the wrong valves and attempting to work on a vessel which is still under pressure has happened, resulting in fatal injuries to the maintenance crew. Use of a mechanical interlocking system will avoid the risk by design. The crew is issued with the keys to allow operation of the correct valves.

5.5.2   Mistaken identity

Another common hazard is working on the wrong equipment. In process plant, it is common practice to overcome this by strict housekeeping to ensure that each instrument and item of equipment has an irremovable, non-corrodible identity tag which is engraved with its unique tag number. This tag number appears on all its documentation and is shown on all design drawings to avoid all possible mistakes. This is of course the reason for colour coding of cables, wires and even pipework.

5.5.3   Reliable isolation

This is important where cross-contamination from other processes can occur. This will be in cases where interconnection is only needed under special circumstances. Cross-contamination could also be a hazard.

More usually, sections of plant need to be isolated for maintenance. Design provisions are needed to avoid the risk of leakage of dangerous fluids into equipment under maintenance. These are all safety issues.

Double-block and bleed valves

These are used in the isolation of equipment, or sections of plant, for maintenance or operation that involves any toxic or flammable gases. By using a double-block valve and vent, anything that leaks across the first valve leaks to a safe location via the vent valve and can be monitored for leaks. This ensures that nothing can pass across into the isolated part.

Spectacle blinds

The provision of spectacle blinds, which are designed to fit between flanges, will provide positive isolation. One disc has a hole the diameter of the pipe that is used for normal operation. It is joined to a second disc, which has no hole. When isolation is needed, the isolation valve is closed. The isolated section can then be made safe and purged of all hazardous fluids. The flange on the safe side of the valve can then be disconnected and the spectacle blind reversed. Reconnecting the flange reassembles it. The disc with a hole is then outside and it indicates that the line is blanked off and safe. This safeguards the isolated section from possible valve leak or inadvertent opening of the valve as the pipe remains blanked off.

5.5.4   Use of full-bore ball valves

Previously, globe valves that were used to discharge corrosive fluids were unreliable due to blockage and corrosion. Drainage of air receivers was a typical example. The water condensed from air is very corrosive and the products of corrosion would accumulate at the bottom of the pressure vessel. Water condensate discharge traps are very unreliable due to debris and corrosion. Substituting these with the use of stainless steel pipework and full-bore ball valves cures this problem as there is nowhere for debris to accumulate. This is a design provision to reduce the risk of corrosion.

5.6 Conclusion

It is hoped that the foregoing has given a sufficient introduction to understanding the complex issues of how to integrate safety into plant and equipment design and how the reliability of systems can be improved.

In the UK, the HSE and the Fire Service can provide assistance in advising on the regulations for fire protection and the means of escape. Plants and buildings will need to be insured and so the insurance companies will also need to be satisfied that the assets being insured will have the risk of fire minimized. In the USA, the OSHA provides advice and issues standards and regulations.

Although the basic issues as outlined in this chapter are universally applicable, in specialist areas such as aircraft and shipping other regulatory bodies will be involved. For example, shipboard fires present a serious hazard to the safety of crew and passengers and the ability to operate reliably. In the last decade these concerns have focused on the need for safety integration as a prime objective. The International Maritime Organization, with the issue and regular updating of Safety of Life at Sea (SOLAS) regulations, are increasingly prescriptive on the design requirements for ships. Many of these requirements are being applied on off-shore installations and there is increasing cross-fertilization on to on-shore installations.

In situations of high risk, as determined by the regulative authority involved, a quantified safety case may be required for the project. There may be a need for consultants to carry out a risk analysis. This will then be a form of safety audit, to confirm that all hazards have been adequately accounted for in the design, and that the provisions to avoid the hazards or to reduce the level of risk are acceptable.

This chapter has shown how identified hazards can be dealt with. In many other cases the hazards present are not easily identified. They may be hidden and must be searched for, which requires the use of more sophisticated procedures.

Chapter 6

Searching for Hazards

6.1 Introduction

The last chapter considered provisions to deal with identified hazards, as found with checklists and the application of safety regulations. Other hazards may be present that require a search technique to find. These range from the application of simple checks to more thorough methods which require the application of a systematic question and answer procedure. The concept of risk ranking is introduced so that the gravity of the hazard can be qualified by its likelihood and consequence.

One technique will be demonstrated by considering a diesel engine and its auxiliaries. The first step will be the construction of a block flow diagram showing all the streams that cross into and out of the diesel engine, as shown in Fig. 6.1.

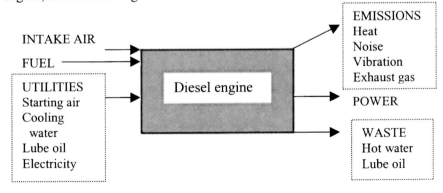

Fig. 6.1 Diesel engine block flow diagram

Each of the streams will need to be examined to find if any hazards are present. Hazards can generally be divided into dangers from the materials used, emissions and energy sources (Table 6.1). Hazards can either be to safety or to health. Wastes and other hazards to the environment are, of course, ultimately hazards to health. Once all hazards are found, a decision can then be made as to what design actions are needed, either to eliminate the hazard or to reduce the risk that could be caused. Of equal importance will be to consider the consequences that could arise from the hazard. If there is an explosion, what other damage could occur and could it have an impact on safety?

Table 6.1 Diesel engine hazards

Hazard	Item	Notes
Material	Diesel fuel	Skin contact can cause dermatitis
	Lube oil	Spillage can cause injury due to slips and falls
	Cooling water	Check safety instruction from water additives manufacturer
Emissions	Exhaust gas	Air pollution, NOX, H_2S
	Noise	Hearing damage
	Vibration	Well-being
	Heat radiation	Dehydration
	Lube oil vapour	Air pollution
Energy	Starting air	Explosion
	Electricity	Shock
	Moving parts	Physical injury
	Diesel fuel	Fire
	Lube oil	Fire
	Hot surfaces	Burns
Waste	Diesel fuel	Sludge disposal
	Lube oil	Lube oil disposal
	Cooling water	Contaminated water disposal

A fire could cause the starting air pressure vessel to explode and the venting down of the vessel will need to be part of the fire protection control system. The consequences of any hazard arising must always be considered. The design action needed will depend on the level of hazard and this will need to be verified by examination of the design data, which are:

Noise emissions	Engine ISO NR 100
	Exhaust ISO NR 130
Exhaust gas temperature	285 °C
Cooling water	Inlet 60 °C, Outlet 90 °C
Starting air	Working pressure, 30 bar max., 8 bar min.
Fuel	Flashpoint 75 °C
Lube oil	Flashpoint 200 °C

Review of the design data confirms that action must be taken on noise and hot surface temperatures. Fire risk from fuel and lube oil will be considered as very low. However, they will feed a fire should a fire occur and if fuel were to spray on to a hot uninsulated exhaust pipe, it will ignite.

There are stringent regulations concerning waste disposal, and this issue will need to be addressed with the authorities concerned.

6.2 Risk ranking (criticality analysis)

In some situations a method is needed to determine how serious a risk the hazard poses. This is done by consideration of the *likelihood* of the hazard occurring, and the *severity* of the consequence. This is a qualitative assessment that depends solely on the judgement of the people involved. It pinpoints areas of risk that require in-depth investigation. The make-up of the risk-ranking matrix has to be decided upon before starting the assessment. As an example, the risk-ranking matrix commonly used in the petrochemical industry is given below. The severity level is selected from the classes as defined in the table. The likelihood of the event occurring is then considered. The risk ranking can be calculated or found from the matrix.

Example

Severity level   Class 3
Likelihood   Class 2

Therefore the risk ranking is 6

6.2.1 Risk ranking matrix

Severity level

Class:		Definition (any one or more):
1	Serious	In-plant fatality; public fatalities; extensive property damage; serious and long-term environmental damage; 2 or more days extended downtime.
2	High	Lost time injury; public injuries or impact; significant property damage; environmental impact exceeding regulation standards; downtime of 1–2 days.
3	Medium	Minor injury; moderate property damage; minimum short-term environmental damage; 4–24 h downtime; disruption of product quality.
4	Low	No worker injuries; minor property damage; no environmental impact; downtime less than 4 h.
5	Minor	No worker injuries, property damage or environmental impact; recoverable operational problem.

Likelihood

Class:		Frequency of occurrence:
1	Frequent	Potential to occur frequently (many times a year).
2	Occasional	Potential to occur occasionally (once a year).
3	Moderate	Potential to occur under unusual circumstances (once or twice in facility lifetime).
4	Unlikely	Could possibly occur, or known to occur within the same industry, but not likely to occur over the facility lifetime.

Ranking matrix

		Severity				
		1	2	3	4	5
	1	1	2	3	4	5
	2	2	4	6	8	10
Likelihood	3	3	8	9	12	15
	4	4	8	12	16	20

Notes
1.	A rank of 1 signifies the most dangerous risk.
2.	The rank of 20 is an acceptable risk.
3.	The shaded area showing rankings from 12 to 20 is usually considered acceptable, needing no action.
4.	Risk ranking is a qualitative assessment that depends on the experience and judgement of the assessor.
5.	The technique is normally used in Failure Mode and Effects Analysis (FMEA) and in HAZOP analysis.

6.3 FMEA

FMEA is a procedure that requires a machine or system to be broken down into subassemblies, or subsystems. Each of the broken-down elements can then be considered in turn to determine the effect of failure on the whole.

When this identifies a subsystem that is critical to the reliability of the whole, then this in turn can be broken down to its components, and the procedure is then repeated.

The technique requires the use of tabular worksheets for completion under headings, which are defined as follows.

Item identity and description
Identification code (useful for a large FMEA where a database may be needed) and a description of the item.

Function
A brief description of the function performed by the item.

Failure modes
As there may be more than one, each failure mode must be listed.

Possible causes
Identify the likely causes of each possible failure mode.

Failure detection method
Design features that could help to detect the failure.

Failure effect
This is subdivided into two subheadings.

Local effect: the effect of the failure on the item's functional performance.

System effect: the effect of the item failure on system operation, plus external consequential damage to other plant.

Compensating provisions
Any internal features of the design that could reduce the effect of the failure identified.

Rank
Carry out risk ranking procedure in accordance with Section 6.2.

Remarks
Record any comments on the failure mode or its effects, including any recommendation for action or design modifications.

Examples of the application of FMEA are given below. Further guidance on criticality analysis and FMEA can be found in reference **(1)**.

6.3.1   A diesel engine FMEA
As an example, an FMEA has been carried out on the utility systems of a marine diesel engine, see Fig. 6.1. Table 6.2 shows the results of an FMEA of the utilities. Not surprisingly nothing critical was found. The loss of an engine at sea will be critical to the safety of the ship. The engine would need to comply with the requirements of one of the classification societies, such as the American Bureau, Lloyd's Register, Bureau Veritas, etc. From the previous work of this chapter it has been identified that the starting air system poses the biggest hazard and so, as an example, the starting air system and its pressure vessel will be studied for possible system failure explosion.

Pressure vessel example
One reason for the starting air system to explode is that the air storage container, a pressure vessel, was wrongly designed or manufactured. To investigate the risk of this, a design and manufacturing process flow block diagram can be constructed. This is shown in Fig. 6.2.

Each one of these elements can be studied for the possible risk of failure. This can be done by the application of an FMEA, see Table 6.3. To do this, however, some knowledge of the design and manufacturing process is required.

At the beginning of the industrial revolution many pressure vessel explosions occurred which led to loss of life. There were problems of faulty design, incorrect material and manufacture, and wrong application. Present-day procedures avoid these problems by the use of internationally recognized design codes such as British Standards, ASME standards and DIN standards, to name just a few. These lay down the working stress based on specified material properties, allowable joint weld efficiency based on required non-destructive testing (NDT) acceptance criteria, and a final hydraulic pressure test to verify structural integrity.

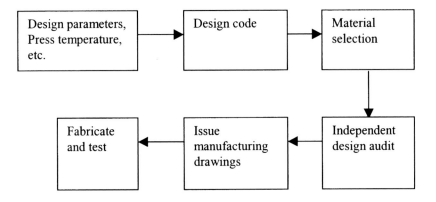

Fig. 6.2 Pressure vessel design and manufacturing process

Table 6.2 Diesel engine FMEA of auxiliaries

Diesel engine auxiliary systems   *Mode: Normal operation*

Item	Function	Failure mode	Failure cause	Failure detection method	Failure effect		Compensating provisions	Rank	Remarks
					Local	*System*			
Starting air	Start-up	Low pressure	Compressor doesn't start	Low pressure alarm (LAP)	Low pressure	Can't start engine	Start spare compressor	15	
Cooling water	Cooling	No flow	Pump fails	LAP	High temp.	Engine overheats	Start spare pump	15	Engine is safeguarded by shut-down
	Cooling	No cooling	Fan fails	High temp. alarm		Lube oil overheats	Spare cooler		
Lube oil	Lubrication	No lube oil	Pump fails	LAP	Low pressure	Hot bearings	Start spare pump	15	
	Cooling	Too hot	Cooling water fails	High temp. alarm	High temp. alarm		See cooling water		
Fuel supply	Combustion	No fuel	Empty tank	Low level alarm	Empty tank	Engine stops	Operating procedure	15	Operator check
Combustion air	Combustion	No air	Filter dirty	Delta pressure alarm	Low pressure	Engine power loss	Trend delta pressure	15	Routine maintenance

Table 6.3 Starting air pressure vessel FMEA

Diesel engine starting air pressure vessel   Mode: Design and manufacture

Item	Function	Failure mode	Failure cause	Failure detection method	Failure effect			Compensating provisions	Rank	Remarks
					Local	System				
Air receiver	Contain pressurized air	Design parameters	Wrong specification	Check data sheet	Highlight errors	Incompatible design	Nameplate		Site inspection
		Design	Error	Third-party audit	Pinpoint errors	Rework required	Design rejected		
		Material	Off spec.	Measure composition and physical properties	Material certs rejected	Unusable material	Order new material		
		Weld design	Wrong weld procedure	Weld procedure qualification	Fail QC	Reject weld procedure	Redesign weld procedure		
		Fabrication	Weld defect	NDT per code requirements	Fail QC	Material reject	Repair weld		
			Fabrication defect	Hydrotest	Leakage or cracks	Item rejected	Scrap and replace		

The risk of misapplication is reduced by the need to apply a non-removable stainless steel metal nameplate. This records the year of manufacture with all the design, working and test pressures, together with other critical data. A third-party inspection stamp is then applied, to certify that all materials, quality control and assurance procedures have been carried out. Usually, annual inspection is required by law to check for development of material defects.

The use of design codes does not preclude the need for designers to consider whether or not degradation needs to be considered in the design. If the design has not allowed for this then unexpected failure could take place; operators and maintenance engineers must always be alert to the symptoms of incipient failure. Typical degradation mechanisms are:

1. fatigue;
2. creep;
3. corrosion;
4. erosion;
5. crack propagation.

These mechanisms should be considered at the design stage. For example, in the case of the starting air vessel, which is subject to cyclic operation, fatigue must be considered in the design calculations. Usually some corrosion allowance is made but this will be inadequate if the air happens to be polluted with H_2S. Another danger will be brittle fracture that may have occurred during transit and storage. Detection of these and other unforeseen hazards will depend on the skill and vigilance of the maintenance engineers in carrying out their routine inspections.

6.3.2   Starting air system pressure control

In the preceding section, the possibility that the air storage pressure vessel might explode due to a defect in its design or manufacture was examined. It was demonstrated that this would be unlikely. Reliability is assured by strict audit and quality control of the design and manufacturing process. This is attained due to quality assurance by: independent design audit, inspection and supply of material; weld procedure; welder qualification; non-destructive test (NDT) records (X-ray, ultrasonic and crack detection); and hydro-test certificates.

With this assurance, any other risk of an explosion can only be due to degradation or incorrect operation in service, resulting in over-pressure. A first concept of a pressure control system could be one where an operator watches a gauge and switches off an air compressor when the maximum pressure has been reached. As required by safety regulations, a pressure safety relief valve further protects the pressure vessel.

Examination of the diagram shows that there are only two ways for the vessel to be subjected to excessive pressure. The pressure safety valve and the operator control work in parallel and are independent of each other. Either of them could stop excessive pressure. The operator, pressure gauge, push button and switchgear are said to work in series. They all depend on each other. If any one fails then they all fail.

The system could be made more reliable by adding automatic pressure control. This has been shown in Figs 6.3 and 6.4 as an addition. With this addition, the system depends on the reliability of the switchgear and the pressure safety valve. The operation of the switchgear now depends on two independent controls (redundancy), one by the operator and the other by the automatic control (diversity). The system is more reliable as more things need to fail before there is excessive pressure.

A logic flow diagram can be made to illustrate the control system. It can be seen that the control logic is operator, pressure gauge, push button, switchgear and compressor. If any one of these elements fails then the whole control system fails. If the control system fails, then the system depends on the reliability of the pressure safety relief valve on the vessel.

The safety of the manual control system can also be examined by the use of FMEA, Table 6.4. It will be seen that the risk of an explosion is unacceptable due to the high risk ranking of 4. The risk is reduced by the addition of an automatic pressure control to the system. This, however, cannot improve the risk ranking because a coarse qualitative assessment cannot assess risk reduction. To assess the reduction in risk a quantitative procedure has to be used. This will be examined in the next chapter.

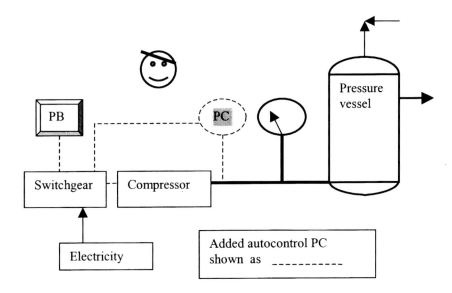

Fig. 6.3 Diagram of manual control system

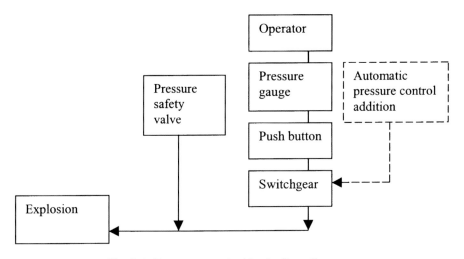

Fig. 6.4 Pressure control logic flow diagram

Table 6.4 Starting air control system FMEA

Diesel engine starting air control system   Mode: Normal operation

Item	Function	Failure mode	Failure cause	Failure detection method	Failure effect		Compensating provisions	Rank	Remarks
					Local	System			
Starting air system	Controls pressurized air	Excess pressure	Operator	None	High pressure	Safety valve opens	Noise of air release	12	Add auto-control
		Ditto	Pressure gauge error	None	Ditto	Ditto	Maintenance schedule	12	Ditto
		Ditto	Push button failure	Operator	Ditto	Ditto	Manual operation of switchgear	12	Ditto, also operator training
		Ditto	Switchgear failure	Operator	Ditto	Ditto	Ditto	12	Operator training
Pressure safety valve	Release excess pressure	Rupture vessel	Safety valve fails to open	Noise	Explosion	Damage to plant and possible fatal injury to operator	Planned maintenance of safety valve	4	In the event that pressure control fails

6.4 Hazard and operability studies (HAZOP)

A HAZOP is a procedure for carrying out a systematic critical examination of an engineering design to assess the hazard potential due to incorrect operation or malfunction of individual items of equipment and the consequential effects on the whole plant. It was conceived as a way of improving safety in the design of chemical plant and is now extensively used in the design of any type of process plant, see reference **(2)**.

A team is needed for the study. It consists of a chairman and a scribe, with representatives from the design team, operations and maintenance.

The actual HAZOP study is a formal review of the process flow diagrams (PFDs) which are conceptual, and piping and instrumentation diagrams (P&IDs) which are detailed designs.

Method

The method requires the design to be divided up into sections, called 'nodes'. For each node, a series of questions called 'guide words' have to be answered. This involves the use of a standard worksheet with specific headings for the answers required.

Procedure

At the start of the study session, the objective of the HAZOP must be stated and a brief background and purpose of the node under study must be discussed. This will enable the team to be focused on the objective. The parameters to be considered must then be decided.

The diagram under study should be displayed on the wall of the study room for all to see. As each line is subjected to the HAZOP, it must then be highlighted, so that at the end of the study it can be seen that all lines have been considered. On completion, the study proceeds to the next node, and so on.

On completion of the HAZOP an initial report is issued, with recommended actions to be taken. A final report is then issued when all recommended actions have been implemented. This becomes an audit and record of what was carried out or, if not carried out, then what was the alternative and why.

HAZOP worksheets

The standard worksheet headings and what they mean, together with the guide words to be used, are listed below. Typical deviations and an explanation of possible causes explain how guide words can be applied.

Worksheet headings

Node	Item or section of plant studied
Guide word	See guide word descriptions
Deviation	Study design and identify meaningful deviations of the guide word
Cause	Identify credible causes of the deviation
Consequence	Assuming that all protection has failed, establish the consequence of the deviation
Safeguard	Identify safeguards provided to prevent deviation
S-Severity	Apply risk-ranking matrix
L-Likelihood	Ditto
R-Ranking	Ditto
Recommendation	Develop recommended action, if needed
Action by	Identify who is responsible to take action

Guide words (and their interpretation)

Guide word:	Typical deviation:	Explanation:
No, None	No flow	Diverted, blockage, closed valve
More	Flow	More pumps, inward leaks
	Pressure	Excess flow, blockage, closed valve
	Temperature	Cooling failure
Less	Flow, pressure	Blocked suction, drain with closed vent
As well as	Contamination	Carry over, inward leaks from valves
Part of	Composition	Wrong composition of materials
Reverse	Flow	Backflow
Other than	Abnormal situations	Failure of services/utilities, fire, flood
	Maintenance	Isolation, venting, purging, draining
	Abnormal operations	Start-up, part load, etc.

6.4.1 HAZOP application example

The example to be studied is based on the starting air system, the concept of which is shown in Fig. 6.3 as discussed previously. However, the air system is to supply utility air for a continuous process plant that must remain in operation for 3 years between shut-downs. In consequence, the air system is to be installed with a spare compressor package and two air storage pressure vessels (receivers). This will allow critical maintenance of the compressors and inspection of the receivers without the need to disrupt the utility air supply. This is a simple example as only one line is involved.

The object of the HAZOP must be to verify safe operation and maintenance without disruption of the air supply. The node under HAZOP study is the air supply to the receivers. The HAZOP is called a coarse HAZOP, as the study will be based on a process flow diagram (PFD).

Findings

The study showed that the closure of any combination of isolating valves would not lead to over-pressure. All sections of pipe up to the receiver isolation valves would be protected by the compressor safety valve. The whole system is of course protected by the pressure control system and the pressure safety valves on the receivers.

It was considered prudent to add an independent automatic high-pressure shut-down and alarm. This will improve reliability at little extra cost. The other recommendation was to add automatic water traps to discharge any water from the receivers and not to rely on the operators. This will reduce the risk of corrosion due to water stagnating in the receiver.

The isolation and venting of the receivers was not provided for. Although inlet isolation valves were shown, the vessel cannot be isolated as the vessel would be pressurized by backflow from the discharge manifold. Although the piping inlet manifold had a pressure gauge, it was considered prudent to add one to each vessel. A pressure gauge on the vessel will enable the pressure in the vessel to be monitored during venting down for maintenance.

Due to the high pressure, all instruments need block and bleed valves to ensure pressure letdown for maintenance.

The HAZOP was carried out on the PFD in Fig. 6.5. The worksheet completed for the study is shown in Table 6.5. The P&ID that embodies the recommendations of the HAZOP study is shown in Fig. 6.6.

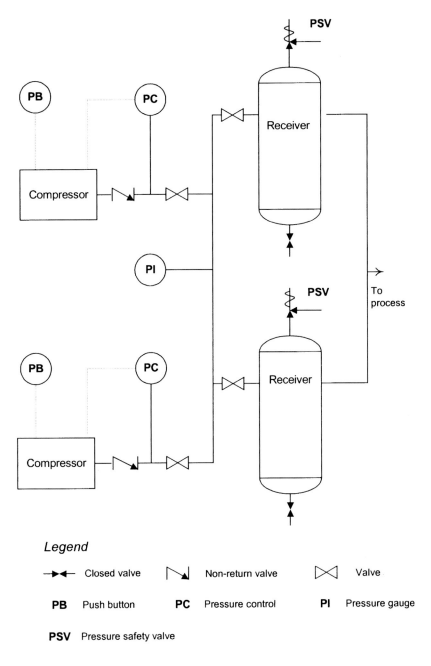

Legend

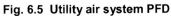

→►◄— Closed valve	Non-return valve	Valve
PB Push button	**PC** Pressure control	**PI** Pressure gauge
PSV Pressure safety valve		

Fig. 6.5 Utility air system PFD

Table 6.5 Utility air system HAZOP worksheet

Session: (date)		Node: Air supply to receivers		Parameter: air flow		Intention: Maintain min./max. pressure	
GW	Deviation	Cause	Consequence	Safeguard	Rank	Recommendation	By
No	No flow	Compressor or receiver valve closed	No air supply	Operator	15	Lock valve in open position	Piping
More	More flow	Excess air supply	Over-pressure	Compressor pressure control	15	Add high-pressure trip as extra safety measure	Design
Less	Less flow	Compressor defect	Lose pressure	Start spare compressor	15	Add to control sequence and alarm operator	Ditto
As well as	Impurity	Moist air	Water in receiver	Operator blow-down	5	Air–water trap	Ditto
Other than	Maintenance	Compressor	Close compressor isolation valve	Permit system	8	Use locked shut valve	Piping
		Receiver	Release air pressure	None	4	Add exit valve, vent valves and pressure gauge	Ditto
		Instruments	Ditto	No vent and isolation valves	6	Add vent and isolation valves	Ditto
More	More pressure	Pressure control fails	System over-pressure	Compressor and receiver safety valve	8	See more flow above	Ditto

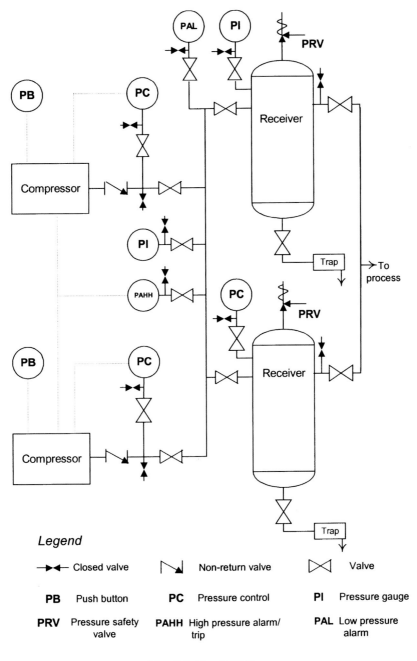

Fig. 6.6 Final P&ID

6.4.2   Other HAZOP applications

The HAZOP procedure was developed by the process industries and the previous example has demonstrated how it can be applied to a P&ID for a process system.

It is also a useful tool for finding weaknesses in any type of system that can be represented by a block flow diagram. It enables the interface parameters to be explored for the effects of any deviation from the planned intent. They could be systems that involve the flow of materials, people or data. Alternatively it could be used in the study of a number of events or activities in a planned sequence. Typical applications are:

- software applications and programmable software systems;
- logistic systems of people and materials;
- assessment of administrative procedures;
- assessment of other systems and devices.

In the HAZOP of logistics where time or sequences are involved, other additional guide words are needed, such as:

- early
- later
- before
- after.

The other guide words may not be applicable and can be ignored.

The IEC standard for hazard studies provides examples illustrating the above applications, see reference **(3)**.

6.5 A cautionary example

The effectiveness of any hazard analysis depends entirely on the experience and creative imagination of the team doing the investigation. The procedures only impose a disciplined structure to the work.

The Concorde supersonic airliner that crashed at Paris in 2000 is a good example of this. During take-off a fuel tank in the wing was ruptured. The escaping fuel was ignited and then the plane caught on fire and crashed. The engineers had considered all failure modes in the design and the fuel tank should not have ruptured. The event that was not foreseen was the possibility that an object could strike the underside of the fuel tank and cause a hydraulic wave to be transmitted to the upper side of the fuel tank. It was the reflected hydraulic wave that then caused the underside of the fuel tank to rupture. If the fuel tank had not been completely full there would not have been a reflected hydraulic wave. For take-off on a long journey the tanks were of course full. No one had thought of this possibility; it just demonstrates how much imagination is needed to ensure that all failure modes are identified. Sometimes it is just too much to expect, as with Concorde.

Making provisions to avoid the hazard by design solved the problem. The tyres were redesigned to avoid bursting and shedding large enough debris to cause damage to the fuel tanks. The fuel tanks were lined with a material that could absorb hydraulic shock waves and self-seal if punctured.

6.6 Summary

This chapter has shown how processes and systems can be broken down and analysed to find hazards to safety and reliability. The techniques of producing block flow diagrams and how to apply FMEA have been demonstrated. A method of risk ranking to qualify risk has been provided.

These methods have been used on an air system, which was developed from an initial PFD to a final P&ID using HAZOP. It has also been shown that finding hazards and reducing risk depend entirely on the abilities of the team assigned. These techniques can be applied to a whole range of situations for many different industries. The work should be a challenge to the creative imagination of any engineer.

In high-risk situations it has also been shown that there will be a need to quantify the risk to safety, and calculate its reliability, for any plant or system. This is especially true if the effects of improvements need to be judged or alternative measures need to be compared. These matters will be dealt with in the chapters that follow.

6.7 References

(1) ISO IEC 60812 (first edition draft revision, second edition 2001) *Analysis techniques for system reliability – A procedure for failure mode and effects analysis (FMEA)*.

(2) *A guide to hazard and operability studies* (1992) Chemical Industries Association, London.

(3) BS IEC 61882 (Due to be published 2001/2002) *Hazard and operability studies (Hazop Studies) – Application Guide*.

Chapter 7

Failure, Statistics and Reliability

7.1 Introduction

The health and safety of workers and the public depends on the reliability of machines and systems. Cities would become a cesspool of disease and death if all the sewage plants, power plants, gas-pumping stations, water-treatment plants and all their distribution systems were to break down. The population would also starve to death if all the transport and communication systems were to fail.

In order to design safe and reliable plant and equipment, it is necessary to obtain and use data. This enables the risk of failure to be quantified so that the data can be used to influence design and maintenance policies.

Statistics is the mathematical science of studying the characteristics of the past in order to predict the future. Engineering safety and reliability depend on predicting the behaviour of machines and components. The life of an item before it fails can only be predicted by the study of past behaviour of similar items in similar circumstances. This depends on the application of statistical analysis. The future has uncertainties caused by different circumstances. This is compounded by the fact that in some cases the numbers studied are often relatively small. This affects the accuracy of prediction so that it can only be regarded as a probable indication.

In situations where the operation conditions are closely controlled, and data can be obtained from a large number of items, the prediction becomes more certain. This is true in the case of automobiles or aircraft. In the case of process machinery, where operating conditions vary and the examples in operation are far less, the data obtained become less reliable.

7.2 The probability of failure, some definitions

Hazard rate
This is called failure rate when it has a constant value. It is usually defined as the number of possible failures in one million hours. Machines and components are considered to exhibit three different types of failure characteristics. These are described below.

The infant mortality characteristic
A newly designed plant or machine has a high failure rate. There are design and/or construction faults that are hidden until it goes into service. When a breakdown (failure) occurs, the fault is discovered and is corrected. The plant or machine is put back into service until the next fault is revealed. As each fault or defect is revealed, more are eliminated. The ones that are left become fewer and fewer until there are none left. At that stage a mature design is then reached. Manufacturers are aware of this, and a new model will be subjected to extended testing before being marketed. This type of failure is more of interest to design and development engineers in bringing new products to market. To a lesser extent, start-up and commissioning engineers are affected during acceptance testing of a new ship or of a new plant.

A mature design
A mature design exhibits a constant failure rate. If something goes wrong, it can be repaired and returned to service with the same failure rate as new. In the design of plant and equipment, it is usual to use proven components of mature design with a constant failure rate.

Wear out/old age
Many engineering components have this characteristic. The longer they are in operation the more likely it is that they will fail. This characteristic is one with an increasing failure rate. Typical examples are valve seats, anti-friction bearings, brakes, tyres and lamps – items that suffer fatigue, corrosion or erosion. These types of failure are the most difficult to deal with as the test results will show a broad scatter of failure rates. The failure distribution takes the form of a bell-shaped curve. There is an average value; they do not fail all at once and there is a broad scatter of results.

The bathtub characteristic

These three phases of life are known as the bathtub characteristic (see Fig. 7.1).

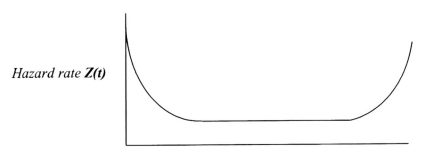

Hazard rate Z(t)

Time t

Fig. 7.1 The bathtub curve

7.3 Statistical analysis, some concepts

To illustrate concepts of failure and failure distribution characteristics, a study of test results from two types of pump will be analysed. In order to do this, some basic terminology must first be established.

Population n

This is the number of items under test in the study.

Failure

It is important to define failure, because there may be degrees of failure. For example, a leak need not be a failure if the pump continues to fulfil its function. However, if the pump were pumping acid, then leakage would also be included in the definition of failure. It may even be necessary to study leakage as a separate failure case. It is important to establish the study objective and define what is to be recorded and analysed.

Time t

This is the time it takes before an item fails.

Raw data

This is the record of the running hours achieved by each of the items of the population studied, ranked in accordance with the running hours achieved.

Class

The total time will be the running hours of the last item to fail, rounded up to the nearest unit time. This time is then divided up into equal class periods, each of which is called a 'class'. As will be seen later, sometimes it may be convenient to combine adjacent classes for ease of manipulation.

Frequency f

This is the number of failures recorded for each class (or period of time), as obtained from the raw data. Often presented as a histogram.

Relative frequency f/n

This is the fraction of the population failing for each class.

Group frequency density distribution f(t)

This is defined as f/n divided by the class interval t, which is the fraction of the population failing per unit time for the class. It is the average value at the mid-point for the class.

Relative frequency density f(t) curve

This is obtained from plotting the values as derived from the group frequency density calculation above. The joining of these obtained points into a curve will point to the type of failure characteristic being experienced. It must be appreciated that the results will only be indicative. Better results depend on a very large population, with small class periods, so that the points obtained are close together. In practice this is rarely achievable with engineering components.

Total fraction failed F(t) (or cumulative distribution function)

This is the f/n at any given class with the summation of the f/n of all the preceding classes. It is the integral of the curve f/n, the proportion failed.

Reliability R(t) (or cumulative distribution function)

The proportion running, plus the proportion failed, must equal 1, therefore $R(t) = 1 - F(t)$ at any point t.

Hazard rate function Z(t)

This is defined as the risk of failure applicable to the items still running at time t. That is, $Z(t) = f(t)/R(t)$, which gives the fraction of the remaining population expected to fail per hour.

Failure characteristics and failure prediction

As illustrated by the bathtub curve, there are three major failure characteristics. By processing raw test data to produce the $f(t)$ curve, the $R(t)$ curve and the $Z(t)$ curve, it will be possible to identify the type of failure experienced. These data can then be used for failure prediction.

7.4 Examples of pump type A

The idealized test results of 100 identical pumps up to the time of failure were recorded. The running time up to failure for each pump is then ranked so that the pump with the least running time is listed first and the one with the longest running time is listed last. The longest running time was 1600 h and so the period of time for the test can be split up into class intervals of 200 h. The test results can then be grouped into classes. In the first 200 h none failed. In the second 200 h two failed, and so on. These results are analysed and given in Table 7.1.

The results from Table 7.1 show that in the second 200 h, two fail and the relative fraction failing is 2 out of 100, which is 0.02. As the class interval is 200 h then the fraction failing per hour is 0.02/200, which is 0.0001.

The relative frequency density fraction failing per hour at the class mid-point, being the average for the class, is plotted and shown in Fig. 7.2. Most of the pumps fail at the time where the curve peaks at about 1000 h as determined by the way in which the curve is drawn.

The data can be looked at in a different way by seeing what percentage or fraction of the total fails with time. After 400 h two have failed and six more fail after 600 h. The total failed after 600 h is eight, which is 8 per cent failed or the cumulative fraction failed of 0.08. Conversely the fraction still running is 0.92. After 1600 h all pumps have failed. These results are shown in Table 7.2; they are shown plotted in Fig. 7.3. This curve shows the fraction of pumps still running as a function of time.

The other result of interest is the probable failure rate, more correctly called the hazard rate. The fraction failing per hour divided by the fraction surviving per hour is the instantaneous hazard rate. Examination of Table 7.1/7.2 shows that the number of pumps failing increases with time so that a greater fraction of the surviving pumps fail with time. This is shown in Table 7.3 and the results are plotted in Fig. 7.4.

7.4.1 Discussion

The bell-shaped curve shown in Fig. 7.2 is typically that for wear out (old age). This type of failure distribution is called a 'normal distribution'.

The average age or mean time between failures, MTBF, is when half of them fail. This is the peak point of the bell-shaped curve in Fig. 7.2. It is also the time at which 50 per cent have failed in Fig. 7.3. The data, presented in different ways, show that an average running life of 1000 h can be expected.

Table 7.1 Pump A failures, processed raw data

Class interval, time to failure	Class mid-point h	Frequency, no. of pumps failing f	Relative frequency, fraction failing f/n	Relative frequency density, fraction failing/h f(t)
0–200	100	0	0	0
200–400	300	2	0.02	0.0001
400–600	500	6	0.06	0.0003
600–800	700	16	0.16	0.0008
800–1000	900	26	0.26	0.0013
1000–1200	1100	27	0.27	0.00135
1200–1400	1300	17	0.17	0.00085
1400–1600	1500	4	0.04	0.0002

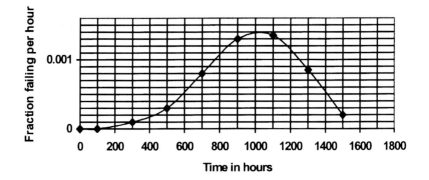

Fig. 7.2 Relative frequency distribution f(t)

Table 7.2 Data for cumulative distribution function F(t), R(t), pump A

Class end	Frequency f	Cumulative cf	F(t)	R(t) = 1 – F(t)
200	0	0	0	1
400	2	2	0.02	0.98
600	6	8	0.08	0.92
800	16	24	0.24	0.76
1000	26	50	0.5	0.5
1200	27	77	0.77	0.23
1400	17	94	0.94	0.06
1600	4	98	0.98	0.02

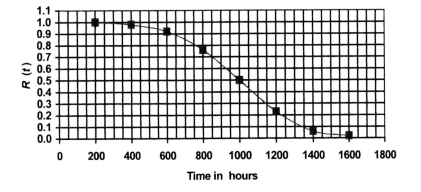

Fig. 7.3 Cumulative distribution function curve R(t), pump A

Table 7.3 Pump A hazard rate calculation

Time (h)	t	300	500	700	900	1100	1300	1500
Fraction failing/h from Table 7.1	f(t)	0.0001	0.0003	0.0008	0.0013	0.00135	0.00085	0.0002
Fraction surviving at t from Fig. 7.3	R(t)	0.99	0.96	0.85	0.64	0.36	0.13	0.02
Hazard rate f(t)/R(t)	Z(t)	0.0001	0.0003	0.0009	0.002	0.003	0.0065	0.0100

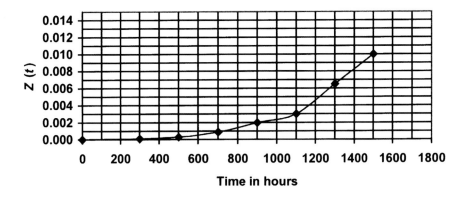

Fig. 7.4 Pump A hazard rate curve Z (t)

The hazard rate curve shows an increase with time, symptomatic of old age, typical of pumps that need to be discarded when worn out. Domestic central heating pumps are an example.

7.4.2 The mean value

The mean value or average value for a normal distribution can be found from the inspection of the curves as shown in Figs 7.2 and 7.3. The mean value can also be calculated.

Taking data from Table 7.1, processed raw data from test results:

the fraction of pumps failing in each class is f/n

the average time for failure at each class interval will be the time up to the class mid-point t

this times the fractional failure for the class will be the average time to failure for that group of failures $t \times f/n$

When these are calculated for each class, and added together, the average time for all failures will be the answer. This is the mean time to failure MTTF. These calculations are shown in Table 7.4. From the table the MTTF is 960 h.

7.4.3   Standard deviation (S)
Examination of the normal distribution will show a scatter of results. A measure of how the actual failures could deviate from the average is called standard deviation.

the deviation from MTTF of the class mid-point is $MTTF - t$

the relative frequency, or fractional deviation of the class mid-point from the MTTF squared is $f/n \, (MTTF - t)^2$

The standard deviation is found by summing the $f/n \, (MTTF - t)^2$ and then finding its square root. These calculations are shown in Table 7.4. This shows the standard deviation to be 264 h, which is a measure of the error of using MTTF. With some other result the standard deviation could be very small. This will be indicated by a relative frequency distribution curve similar to Fig. 7.2 with more of a spike-shaped curve.

The calculated MTTF of 960 h compares well with the value as read from Fig. 7.2 showing a peak at about 1000 h. Figure 7.3 shows that 50 per cent of the pumps fail at 1000 h.

If the standard deviation S is 264, then referring to Fig. 7.2 it can be seen that:

960 + 264 = 1224 h where 83 per cent has failed
960 − 264 = 696 h where the reliability is 85 per cent

Table 7.4 Calculation for the MTTF and the standard deviation

Mid-point t	f/n (from Table 7.1)	t x f/n	(MTTF− t)	f/n (MTTF− t)²
100	0	0	860	0
300	0.02	6	660	8712
500	0.06	30	460	12 696
700	0.16	112	260	10 816
900	0.26	234	60	936
1100	0.27	297	140	5292
1300	0.17	221	340	19 652
1500	0.04	60	540	11 664
	MTTF =	960	S^2 =	69 768
			S =	264

7.4.4 Conclusion

For items with normal distribution failure characteristics, it is impossible to use the mean failure rate as a basis for maintenance. The resulting reliability would only be 50 per cent. For the pump studied, for 98 per cent reliable operation, items would need to be discarded after 400 h; this would be uneconomic, because for half of the time the items would have some 600 h operating life left. Ideally they should be used where the consequences of failure are non-critical and only discarded when they fail.

7.5 Examples of pump type B

These are idealized test results of 1000 identical pumps of a mature design. In this example the total running time was 13 000 h and the adopted class interval is 1000 h. Otherwise the definitions and procedures used are the same as for the analysis of pump A.

Table 7.5 Pump B failures, processed raw data

Class interval time to failure in hours $\times 10^3$	Class mid-point in hours	Frequency, no. of pumps failing f	Relative frequency fraction failing f/n	Relative frequency density, fraction failing/h f (t) x 10^{-4}
0–1	500	300	0.3	3
1–2	1500	210	0.21	2.1
2–3	2500	150	0.15	1.5
3–4	3500	100	0.1	1.0
4–5	4500	70	0.07	0.7
5–6	5500	50	0.05	0.5
6–7	6500	36	0.036	0.36
7–8	7500	25	0.025	0.25
8–9	8500	18	0.018	0.18
9–13	11 000	30	0.03	0.075

Notes
1. Pumps were taken out of service when they failed.
2. The test was discontinued after 989 pumps had failed.
3. The last class time period was increased from 10 000 to 40 000 h in order to accommodate the longer running hours to failure of the last pumps. The actual time period for each class has to be used in calculating $f(t)$.

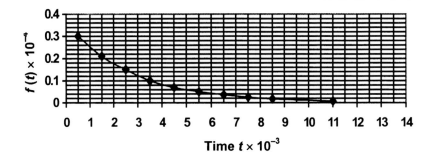

Fig. 7.5 Relative frequency distribution $f(t)$, pump B

Table 7.6 Data for cumulative distribution function $F(t)$, $R(t)$, pump B

Class end	Frequency f	Cumulative cf	F(t)	R(t) = 1 – F(t)
1000	300	300	0.3	0.7
2000	210	510	0.51	0.49
3000	150	660	0.66	0.34
4000	100	760	0.76	0.24
5000	70	830	0.83	0.17
6000	50	880	0.88	0.12
7000	36	916	0.916	0.084
8000	25	941	0.941	0.059
9000	18	959	0.959	0.041
13 000	30	989	0.989	0.011

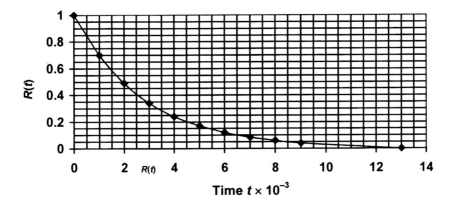

Fig. 7.6 Cumulative distribution function curve R(t), pump B

Table 7.7 Pump hazard rate Z(t) pump B

Time t (h)	Fraction failing/h, see Table7.5, f(t) $\times 10^{-4}$	Fraction surviving (Fig. 7.6) R (t)	Z(t) = f(t)/R(t)
500	3	0.85	3.5
1.5×10^{3}	2.1	0.58	3.6
2.5×10^{3}	1.5	0.42	3.57
3.5×10^{3}	1.0	0.28	3.57
4.5×10^{3}	0.7	0.2	3.5
5.5×10^{3}	0.5	0.15	3.3
6.5×10^{3}	0.36	0.1	3.6
7.5×10^{3}	0.25	0.07	3.57
11×10^{3}	0.075	0.022	3.4

Note
$Z(t)$ is 10^{-4}.

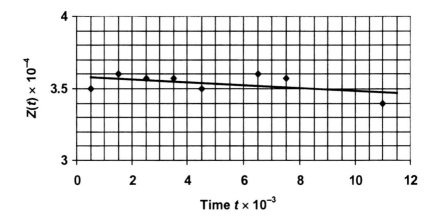

Fig. 7.7 Pump B, hazard rate curve

For reliability engineering, where concern is for the reliability of an item in the future, examination of the fractional failure (cumulative distribution function) curve is required.

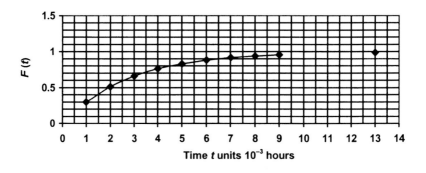

Fig. 7.8 Cumulative distribution function curve $F(t)$, pump B

7.5.1 Discussion

The curve $F(t)$ is asymptotic to 1, decays exponentially and is of the form:

$$F(t) = 1 - \exp(-\lambda t)$$

If the curve is assumed to represent a single pump then the probability of the pump failing is

$$P = 1 - \exp(-\lambda t)$$

at any time t.

Examination of the curves $f(t)$ and $Z(t)$ shows that they tend to intersect the Y axis at 0.356; this is the value of λ for the type of pump which was tested. From Fig. 7.8 it can be seen that $P = 0.3$ when $t = 1000$ h if these values are substituted into the equation of the curve, that is

$$P = 1 - \exp(-\lambda t)$$

then

$$P = 0.3 = 1 - \exp(-\lambda t)$$

rearranging

$$0.7 = \exp(-\lambda t)$$

therefore

$$\ln 0.7 = -\lambda t \quad \text{so that} -0.356 = -\lambda t$$

as

$$t = 1000$$

Then λ has the value of 356×10^{-6} which confirms the finding from the examination of the curves $f(t)$ and $Z(t)$.

Having found the value of λ, the equation of the curve has been established and so it is possible to find the value of P at any time t.

However, as the numerical value of λt is usually less than 1, the equation can be simplified.

The series expansion of $\exp(-\lambda t)$ is

$$1 + (-\lambda t) + [(-\lambda t)^2/2!] + [(-\lambda t)^3)/3!] \text{ etc.}$$

When λt is very small – much less than 1 – the powers of λt can be disregarded as being insignificant, therefore substituting

$$P = 1 - [1 + (-\lambda t)] \quad \text{and} \quad P = \lambda t$$

Assuming that the test results can be taken to be characteristic for all identical pumps, then P is the probability of any identical pump failing within the running hours time t.

As noted above, while the failure rate remains constant the probability of failure will increase as time goes on. If the pump is newly installed and time t is 1, then $1/\lambda$ is the MTBF. For the pumps as tested, the MTBF is $1/356 \times 10^{-6}$, which is 2809 h.

7.5.2   Conclusion

As will be shown later, most engineering systems are made up of components that exhibit a constant failure rate. It is usually also assumed that when they fail they can be repaired as good as new. This enables the probable reliability of such systems to be evaluated. The reliability of such results may be doubtful but it does enable different systems to be compared and improvements quantified, as will be demonstrated in the following chapters.

7.6 Weibull analysis, reference (1)

7.6.1   Background

The foregoing is a simple introduction to the rather complex subject of data gathering and analysis. The bell-shaped curve for normal distribution is used for sampling and the width of the bell shape indicates the deviation from the mean, which is usually the desired objective. The properties of this curve and the ways of defining it will be covered in any book on statistics.

The bell-shaped curve and the other two types of curves (infant mortality with decreasing failure rate and maturity with constant failure rate) are only the basic types. There are many variations for which it would be difficult to find equations. The Weibull analysis procedure is an alternative way to deal with test data. It is a graphical procedure that avoids the use of calculations.

7.6.2 Weibull analysis

The procedure for gathering raw data and then producing a probability density function curve $f(t)$, and a cumulative distribution function curve $F(t)$ is straightforward, as already shown. The real challenge is to identify the curve characteristic equation for $F(t)$. To solve this problem, Weibull proposed a general equation that could fit any curve shape.

$$P = 1 - \exp\{- [(t - \gamma)/\eta]^{\beta}\}$$

The problem is how to evaluate, and find the variables that make up, the function which determines the index of the exponential. This is solved by the use of Weibull distribution function graph paper, available from Chartwell.

- P or $F(t)$ is from the test data at time t.
- η is the value of t when $F(t)$ is about 0.63. This is specially marked on the special graph paper and is called the characteristic life.
- In most cases γ is assumed to be zero; it is a correction factor to be applied afterwards in special circumstances and can be ignored for the examples under discussion.
- β is read off a scale on the graph paper, after plotting the test data, and then completing the required construction. This is called the shape factor because it indicates the type of distribution curve. These are:
  - $\beta = 1$ is for an exponential distribution, where the hazard rate is constant.
  - $\beta > 1$, there is an increasing hazard rate with time. The greater the value the greater the increase (normal distribution).
  - $\beta < 1$, there is a diminishing hazard rate with time.

The results of finding the indices will enable the reliability and the MTBF data to be extracted. This will be demonstrated by using the Weibull procedure on the pump A and B test results.

7.6.3 Pump A

The processed raw data from the pump test as given in Table 7.2 are plotted on the Weibull distribution function graph paper. It should be noted that the value in Table 7.2 for $F(t)$ is given as a fraction and must be converted to a percentage as stated on the Y axis of the graph paper.

The results are shown in Fig. 7.9. It can be seen that the data, when plotted, are a straight line. A reference line must also be constructed from the estimation point, which is marked at the top left-hand corner of the graph paper. The reference line must be drawn so as to be perpendicular to the plot of the data results.

Information from the Weibull construction

1. The reference line intersects the $P\mu$ scale at 49; this means that the time at the cumulative percentage failure of 49 per cent for the plotted data line gives the MTTF. This shows a time of 1000 h. This can be compared with the previous calculated result of 960 h from Table 7.4 or the results from Figs 7.2 and 7.3, also of 1000 h.
2. The graph shows a dotted line at 63 per cent, which is the expected failure time for 63 per cent of the pumps. This is the characteristic life η and the time of 1100 h can be read off the graph.
3. The reference line intersects β at the value of 4. This means that the pump failure has a normal distribution characteristic (value > 1).

Standard deviation

The Weibull method also provides a way of finding standard deviation. The value of β (4) is entered in the chart (Fig. 7.10) and the value of B can be read off as 0.25.

Standard deviation $S = \mu \times B$ which is $1000 \times 0.25 = 250$ h

This compares with the value of 264 h as previously calculated.

Finding the equation for F(t)

By substituting the indices found, the curve characteristic equation is

$$P = 1 - \exp\left[-(t/1100)^4\right]$$

Substituting 1000 h for t gives 0.49, which is 49 per cent, the same as from the plotted data.

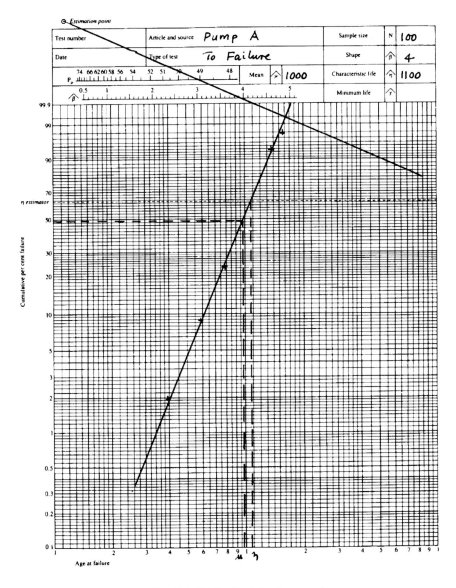

Fig. 7.9 Weibull plot of pump A

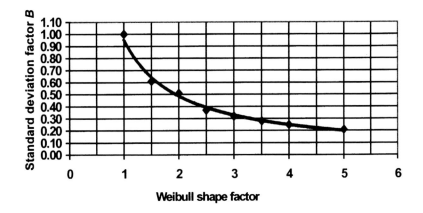

Fig. 7.10 Standard deviation factor B

7.6.4 Pump B

In the same way as for pump A, the data from Table 7.6 can be used for plotting on Weibull distribution function graph paper. The results are shown in Fig. 7.11.

Information from the Weibull construction

1. The reference line intersects the $P\mu$ scale at 63; this means that the time at the cumulative percentage failure of 63 per cent for the plotted data line gives the MTTF. This shows a time of 2800 h. This can be compared with the previous calculated result of 2800 h as derived from the value of $1/\lambda$.

2. The graph shows a dotted line at 63 per cent, which is the expected failure time for 63 per cent of the pumps. This is the characteristic life η and the time of 2800 h can be read off the graph.

3. The reference line intersects β at the value of 1. This means that the pump has a constant failure rate.

4. The Weibull equation as given above can now be evaluated as follows

$$P = 1 - \exp\{-[(t-\gamma)/\eta]^{\beta}\}$$

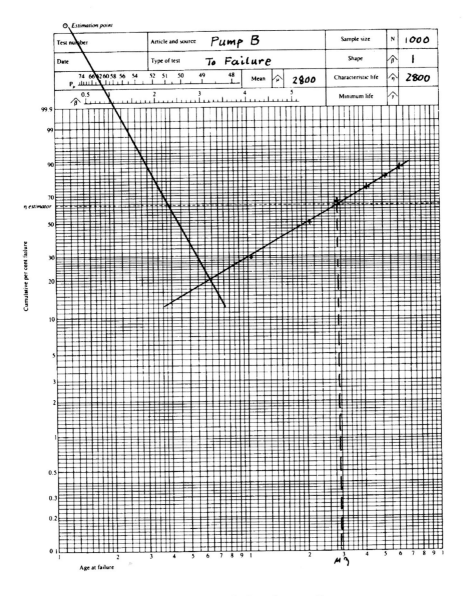

Fig. 7.11 Weibull plot of pump B

As γ is not applicable and β is 1, the equation becomes

$$P = 1 - \exp[-(t/\eta)]$$

This equation is the same as the equation previously discussed for exponential failure distribution, and $1/\eta$ is the same as λ and, of course, as

$$1/\eta = 1/2800 = 357 \times 10^6$$

which is the same as that found before.

7.6.5   Conclusion

Sufficient has been shown to demonstrate the power of the basic way that the Weibull analysis can be used. The two standard cases have been demonstrated for values of β, and the following should be noted.

1. If the results are not in a straight line then it is very possible that two types of failure are mixed together. The data will need to be screened and the data for different types of failure separated for plotting.
2. There may also be distortions that are caused by the location of a bell-type normal distribution.

These and other interpretations on Weibull plot results may be found in other advanced texts on the subject.

7.7 Designing for reliability

It has been shown that reliability can be measured by failure rate, MTBF and MTTR. These are probabilities, not absolute values. It is therefore possible to design for reliability using a systematic approach that will improve these probabilities in the direction of the required reliability. The first step is to establish what is required.

7.7.1   The reliability specification

This must be based on the business goals of the user.

Example

A bus company will require reliable buses that will transport passengers without breakdown with a service life of say 10 years. Investigation will reveal that the daily period of operation may only be 10 h and that not all the buses are used on Sundays. From the study of operating patterns it should be possible to establish agreement for:

- life cycle cost
- an acceptable failure rate
- MTBF
- MTTR.

These figures will be dictated by the need to minimize maintenance costs. Reliability in service is assured if the condition of critical parts could be monitored so that maintenance work is limited to the times when the buses are out of service and the MTTR is less than the time available. These data then allow the target planned maintenance and inspection schedules to be established.

7.7.2 Design analysis

The product design can then be broken down to its subassemblies and components. Data must then be gathered for each item, concerning failure rates, failure modes, wear out characteristics and material properties, etc. These data can be sourced from in-house experience such as:

- manufacturing QA/QC data;
- functional test and acceptance test data;
- in-service data.

Data can also be obtained for items not previously used in-house but that have been used by others, from data books or component manufacturers.

7.7.3 Reliability analysis

This can go from top to bottom by first considering the whole machine and then its subassemblies. FMEA can be used to identify critical failure modes that affect reliability. Fault tree analysis (FTA), as discussed in the next chapter, can be used to estimate failure rates. These techniques will identify critical assemblies and components. Where previously utilized subassemblies have been adopted in the new design then their function must be reviewed to check if there are any variations in their use that could affect its reliability. If there are any variations then a component-by-component review will be necessary. Each component must be tabulated to verify that each one is still being used in the same way as previously and that none are subject to any extrapolation that could reduce its reliability.

Component reliability, however, is a function of its design. Items that exhibit a normal distribution failure characteristic, such as given in the example of pump A, can be said to be unreliable. Ideally the bell-shape curve as shown in Fig. 7.2 should be more of a spike. The fraction failing would be grouped close to the mean so that the period of failure-free operation can be relied on with confidence. To achieve this requires each failure mode to be identified and the deviations that lead to their inconsistent failure rate to be eliminated.

In the design of a customer-specified item, such as a pressure vessel, reliability of design is secured by well-established methods embodied in codes. These require strict control of material properties, fabrication quality control procedures and control of its use.

In the case of items in more common use where such controls are unusual, failure can often be experienced due to the uncertainty of factors such as the following.

Load conditions

This is often wrongly specified or assumed and is one of the primary causes of failure. The product will very often be used in a way, and under conditions, that were not visualized in the design process. Those that operate as designed will be more reliable than those that are subject to rough treatment. The maximum load stress that each of a large population of similar items could experience in use will display a normal distribution (bell-shaped) curve. Most will operate near design; some will experience lesser loads and some more.

Material properties as used in the design

Material specifications also provide specified physical properties. The results of research by the American Society of Metals indicated that there could be a normal distribution for such values as yield strength and ultimate tensile strength (UTS) in the case of mass-produced materials. The physical properties of each item from a large population of the same component will not be the same. They have been found to show a normal distribution.

The load–strength interference concept

In design the above uncertainties are taken care of by the use of a safety factor (SF). In a new design the problem is what value of safety factor to adopt. The safety factor is usually defined as

SF = UTS/load stress

Due to the uncertainties of the load and material properties the idea of safety margin (SM) is more appropriate

SM = UTS – load stress

This is illustrated by the load–strength interference diagram, Fig. 7.12.

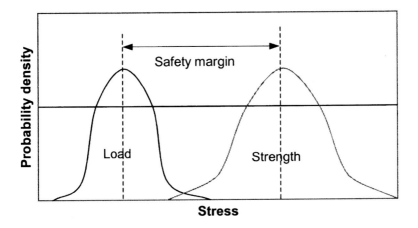

Fig. 7.12 Load–strength interference diagram

The chart demonstrates that the adopted safety margin is not enough. When the possible variations in load and strength of the materials used are shown, the two distribution curves cross over. This shows that it is possible for some components to have a low strength, and if they happened to be used in situations where higher-than-normal loads are experienced then they will fail.

Even if the safety margin is adequate when new, degradation of material physical properties can occur due to corrosion, erosion, creep, etc. Conversely, loads could change due to the degradation of other related components.

Environmental conditions

Most wearing components fail early due to an unsuitable working environment. Reliability can be dependent on design features to control their working conditions.

Antifriction bearings are sensitive to debris. Any foreign matter entering the bearing area can cause rapid failure. Reliability then depends on the effectiveness and reliability of its sealing system. This is a function of design.

Process pumps can be unreliable due to premature seal failure. Seals will fail if foreign matter enters into the seal area or if they overheat due to the lack of cooling. API standards **(2)** recognize this and an array of methods for maintaining the seal environment are provided for the designer to choose.

Process reciprocating compressors use plastic rings. The wear rate is very much affected by the operating temperature. The operating temperature is affected by the compression ratio for each stage. In service, valve deterioration will cause a change in compression ratio, which then affects the temperature. Design features to control this will maintain piston ring life and avoid premature maintenance shut down.

In other situations the working environment is outside the control of the designer and is due to the operator. Wear rate due to erosion, corrosion, fatigue, creep and fracture are affected by working conditions. Early failure is due to unforeseen conditions that need to be identified in service for subsequent modification to the design. This is in-service data feedback that is required for reliability development.

7.7.4   Reliability assessment

After all the analysis work has been completed, an assessment is needed to verify whether the reliability specification can be met, and if not, what action is needed to meet it. The outcome can be categorized as follows:

- items that have to be replaced;
- items that can be repaired;
- items whose reliability is not acceptable and their failure modes.

In order to meet the reliability target for the whole machine, there will be a required reliability target for each item. All the constituent items of the machine must match the scheduled inspection and maintenance period required for the whole machine. Items can be complete subassemblies or components.

Perhaps materials will need to be reviewed or design loads adjusted or different technologies required. Maybe redundancy and/or diversity needs to be introduced. The best economic, reliable design requires the optimum balance of items, and needs mathematical analysis of:

- which items should last the expected life of the machine;
- which items to replace and their useful life;
- which items to repair and their service duration.

In engineering it is important to maximize the use of proven components and limit the new features to be adopted. This enables new products to be launched after development and testing centred on the new features. In the design of any new product the engineer has to be conscious of the limitations that have to be faced. Engineers must be prepared to conclude that in some cases the reliability specification cannot be met – some compromise may be needed or the project should be abandoned. Many projects fail because engineers have tried to advance beyond proven or available technology. For example, the need for mass passenger air transport was foreseen in the 1950s. Attempts were made to design and build these aircraft. They needed long runways and were built with eight engines. They were unreliable and were a failure because engines of the correct power were not available. The prototype crashed during development testing. It took another 10 years or so before the jumbo jets went into service.

7.7.5 Reliability development

As previously shown, the introduction of a new product normally has a high failure rate. Extensive prototype testing may be needed to identify these failure modes before going into production. In other cases machines may be sold to selected customers for operational use in applications where reliability is less critical.

Many operational failures are due to operating conditions that are not specified or foreseen by the designer. Reliability improvement is then dependent on in-service data feedback. A schedule of reliability reassessment is then needed for the review of failures experienced, inspection reports and maintenance activities, in order to identify modifications that may be needed and any research or testing requirements. A regular review of maintenance and inspection schedules must also take place to account for these findings.

Development of items for reliability is a long and expensive business. Different operators will have different operating conditions, which result in the need for obtaining data from many samples over a long period of time. Periods between maintenance and inspection can only be increased with caution, based on solid operating experience and the resolution of the failure modes encountered.

Problems of reliability development

In the example of the Apollo mission to the moon, it took 10 years to ensure the reliability of a mission duration of a few weeks.

In the case of capital plant such as power stations and petrochemical process plant, an installed life of 20 years is normally expected. Non-stop operation of 4–5 years is required with a short shut-down of a few weeks. The expertise of the maintenance engineers and supporting specialists with adequate data feedback is critical in developing reliability.

In a new design, to ensure reliable operation for 4 years requires testing maybe for 10–15 years. Oil companies commonly will not accept a new design for operation in a plant unless there is a proven track record with three to four examples in reliable operation for more than 5 years.

In many cases reliability improvement can only come about through the development of new materials and technologies. Consider the development of aircraft and automobiles and the reliability improvements over the last few decades. These have been dependent on the application of computer modelling and improved materials, which were not previously available.

Unfortunately, reliability can only be established through operation and testing. There is no short cut.

7.8 Summary

This chapter has provided an introduction to data gathering and analysis of equipment failure. It establishes the basic equations that can be used in reliability analysis and the quantification of risk.

Reliable operation is achieved when in-service failures of equipment and systems are avoided. Failure of any component of an aircraft in flight could lead to disaster. There is a need to replace all critical items before they fail so as to ensure safe operation. It has been shown that the use of MTTF to decide on when to change items is dangerous for safety critical items. Due to the wide deviation possible, to ensure safety many items would need to be taken out of service well before they fail. It is not cost effective. It has been shown that such items need to be designed to be more reliable. The role of statistics in design has also been introduced, as has the need to ensure reliability by considering all the possible variables in operation that the product will experience. The risk of failure in new projects and the dangers of exceeding the boundaries of developed materials and technology have been discussed.

As will be explored, one solution to improving reliability is the application of the principle of redundancy and diversity. The other will be to measure the effects of age or the onset of failure, which is condition monitoring.

The role of data gathering and statistical analysis only provides an understanding of the time it takes to fail. Engineers need this data to identify where to direct their efforts in finding solutions to the problems of improving safety and reliability.

The next chapter will show how failure data can be used in the quantification of risk.

7.9 References

(1) **Davidson, J.** (1989) *The reliability of mechanical systems* IMechE, ISBN 0 85298 881 8.
(2) API standard 610 Seventh edition (1989) *Centrifugal pumps for refinery service.*

Chapter 8

Quantifying Risk

8.1 Introduction

To comply with health and safety law engineers need to design things so that the inherent risk is as low as reasonably practicable (ALARP). Therefore it will be necessary to understand and examine how to interpret 'as low as reasonably practicable'.

As demonstrated in Chapter 6, it is possible to apply a risk-ranking technique to give a qualitative judgement on what is a higher or lower level of risk. In the case of preventing a pressure vessel explosion, as shown, its risk ranking is not reduced by adding more safety features. For these sorts of risks, a way of quantifying them is needed in order to judge the effect of adding more safety features. To do this, failure data have to be obtained for all the common engineering components that make up a system. The way in which these data are to be used must be established, and then a procedure to analyse and evaluate the possible failure rate of the whole system.

8.2 ALARP

In Europe, in contrast to the USA, the legal words 'so far as is reasonably practicable' were deliberately chosen to be open-ended so that any and every situation is covered without the need for prescriptive regulations. If an accident occurs, it then has to be proven that steps had been taken to prevent the accident by reducing the risk from the hazard to 'as low as reasonably practicable'.

Example of ALARP

For a building that requires roof maintenance access, the following alternative facilities to be provided can be considered:

1. permanent stairway up to the roof with hoist facilities;
2. permanent external wall ladders with intermediate platforms and hoist facilities;
3. no facilities, use contracted mobile equipment when needed;
4. leave it to the owner's maintenance department.

The choice made will depend on a number of factors:

- The first option will have the highest cost, and each following option will cost less. How much money must be justified?
- The cost then has to be balanced against the risk of a man falling.
- The risk of falling depends on how often there is need to go on the roof.

If there is a need to go on the roof only once in every 5 years, it clearly is not reasonable to insist on the expense of the first two options. Once a year, perhaps, could justify option 2, and perhaps once every few weeks, option 1. Option 4, which implies no preconsidered plan, would be against the law.

8.2.1 Acceptable probable risk

It is generally considered that risks which have a fatal injury (hazard) rate of 10×10^{-5} or more are unacceptable. One in a million is considered safe. Table 8.1 shows the published figures for fatal injury rates for various industries and other activities, compared with the risk-ranking limits given above, (1, 2).

Table 8.1 Comparison of fatal injury rates with risk acceptance criteria

Activity	Fatal injury rate per 10^{-5} persons per year	Risk acceptance criteria
Heavy smoking	500	Unacceptable
Rock climbing	400	Ditto
Mining	100	Ditto
Road user	10	Only just tolerable. But can the risk be justified?
Agriculture, hunting, forestry and fishing	7.5	Tolerable, but needs justification
Construction	4.7	Ditto
Extraction and utility supply	3.2	Ditto
Manufacturing	1	Ditto
Services	0.4	Tolerable
	0.1	Acceptable
Lightning	0.01	

It is interesting to note that, as a design criterion, the level of risk from voluntary activities is considered unacceptable for workers and to the general public. This must be so, because people at work, or using public facilities, do not expect to be exposed to any significant danger.

Safety costs money and therefore it has been recognized that to achieve a very low risk may not be economically viable. A judgement then has to be made on how far the risk can be reduced by measures that are 'reasonable and practicable'. In some situations it may even be too risky to proceed. In other situations, such as mining, people accept the high risk.

8.3 Failure rate application

Engineering systems, which are designed to be as reliable as possible, only make use of well-proven components of mature design. These components will have constant failure rates and this is characterized by the equation (where $\lambda t \ll 1$ as explained in the previous chapter)

$$P = \lambda t$$

As can be seen, the probability of failure P increases with time t. If the pumps were to be repaired as soon as they failed and put back into service, and they were assumed to be as good as new, then the failure rate would always be constant. Based on these assumptions, the equation can then be applied in two different ways.

8.3.1   Revealed failures

These are failures that are made known as soon as they occur. In a ship, which is cruising along, engine failure will immediately be noticed and the engine will remain stopped until the defect is corrected. In these situations it is usual to take t to be the mean time to repair (MTTR). In this case, λt is the fractional deadtime or downtime. It should be noted that the MTTR is the total downtime, which includes administration of permit to work systems, and waiting for labour, spare parts and any testing needed before putting back into service. The MTTR, therefore, will vary from organization to organization and must be specific for any given situation.

8.3.2   Hidden failures (failure on demand)

A firewater pump is not normally in operation; it is only in use when there is a fire. It is therefore possible that the pump is in a failed state that is only revealed when it is called upon to operate. This is a hidden failure or failure on demand. Therefore, when equipment is idle, it is necessary to test at regular intervals in order to verify that it is in working order. When on occasion it is found to be defective, it can be assumed that, on average, it must have been in a failed state half-way between testing. In this type of situation it is usual to use $T/2$ in place of t, where T is the testing interval. As can be seen, the selection of the testing interval affects the failure rate and, as will be demonstrated later, will affect the overall safety of any given system.

8.4 Equipment failure rate

The previous chapter gave an example for finding the failure rate of a pump that was constant. In the same way it will be necessary to establish the failure rates of a whole range of equipment in order to be able to model systems and find their probability of failure. This work has been carried out due to the need to ensure safety in the nuclear, oil and gas, and chemical industries, and data books are available with failure rate data.

Table 8.2 Generic equipment failure rates, reference (3)

Item	Failure rate λ/M	MTTR t (h)	Test interval T/2 (h)	Probable failure rate P/M	Comment
Human operator				350	$M = 10^6$
Pushbutton	0.2	4	–	0.8	
Pressure/temperature gauge	11	3	–	33	
PAHH S/D switch	8	–	1000/2	4000	
Auto-pressure control	12	8	–	72	
Pressure relief valve	0.26		500/2	65	Fail to open
Switchgear	8	60	–	480	
Electric motor	5	60	–	300	
Electric supply	50	1	–	50	
Compressor	400	100	–	40 000	
Diesel engine, small	3250	84	–	273 000	i.e. 0.273
Diesel engine, large	1278	100	–	127 800	i.e. 0.1278

Table 8.2 gives some typical values of failure rates and probable failure rates. These will be used for the calculation of system probable failure rates in the examples that follow. It should also be noted the values assigned to MTTR and the testing interval are purely arbitrary.

8.4.1 Human failure rate
This is covered in Chapter 4, Section 4.6.

8.5 Basic concepts

In this section some basic building blocks for risk quantification will be presented.

8.5.1 Series systems
A manual control system, which consists of an operator, a gauge, a pushbutton and switchgear, is a series system. There are four elements and failure of *any one* will cause the system to fail. This can be represented by a block flow diagram, where P is the probability of failure.

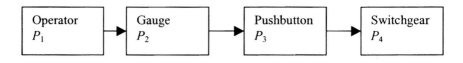

Fig. 8.1 Series block flow diagram

It can also be said that the failure of P_1 or P_2 or P_3 or P_4 would cause failure. This system can also be shown as a logic diagram.

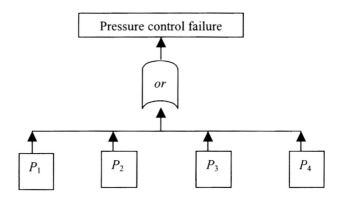

Fig. 8.2 Series logic block diagram

Four ways in which failure can occur means that there are more chances of failure. This means that the probable failure has to be greater than any one of them individually. Therefore for series systems the probabilities must be summed. A series system is less reliable and more prone to failure.

Series (*or* gate) sum: $P_{\text{system}} = P_1 + P_2 + P_3 + P_4$

8.5.2 *Parallel systems*

This is the mathematical expression of *redundancy*, when there is more than one way of fulfilling a function. For example, a man has four vans at his disposal and has an urgent delivery. If one fails to start, he has three others to try. He has 300 per cent redundancy. They must all fail before he is unable to go. The probability of failure must be less than for only one van.

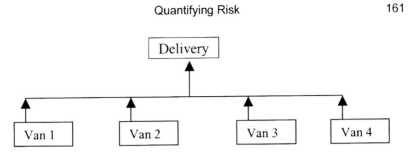

Fig. 8.3 Parallel block flow diagram

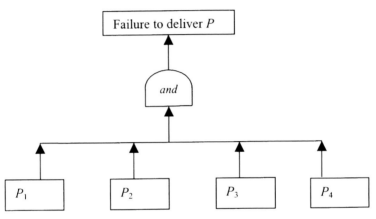

Fig. 8.4 Parallel logic block diagram

The probability for failure for a parallel system is

Parallel (*and* gate) multiply: $P_{\text{system}} = P_1 \times P_2 \times P_3 \times P_4$

Therefore a parallel system is more reliable and less prone to failure. Note this is on the assumption that P has a failure rate of less than 1.

8.5.3 *Reliability*

As stated above, if $P = \lambda t$ when t is the MTTR, then P can be said to be the fractional downtime. The fractional up-time, or its availability, must then be $1 - P$. If it is up and running, then it is known as being available for use and therefore can be relied upon. Therefore the equation for reliability is

$R = 1 - P$

8.5.4 Combinations

The preceding section showed how to evaluate parallel systems. For the example given, only one van was needed and there were three spare vans available (300 per cent redundancy).

On another day there could be a different situation. Three vans are in constant use and there is one van held as a spare (33.3 per cent redundancy). Because all the vans are identical, which ones are used is of no concern. Van A, B and C is no different to van C, B and A. The different vans available are a combination, and not a permutation. In order to calculate the probable failure to deliver, use must be made of the binomial distribution equation. This is developed as follows.

Combination of vans

- Number of vans A B C D
- For the system to fail, any two vans must break down. These failure modes are:
  Combination 1: AB, AC, AD
  Combination 2: BC, BD
  Combination 3: CD

By examination, it can be seen that there are six possible combinations of two vans failing that can cause delivery failure. The chance that there are only two vans depends on probability of failure of any two vans *and* the reliability of the other remaining two vans to operate. That is

$$P^2 \times (1 - P)^2$$

This is for any one combination and, as there are six combinations, then the probability for any two vans to fail will be

$$6 \times P^2 \times (1 - P)^2$$

The general equation for a binomial distribution, which caters for any number of combinations, is

$$P \text{ of system} = \{n! / [r!(n - r)!]\}\ P^r (1 - P)^{(n - r)}$$

where
  n is the number of items available, 4 in the example
  r is the number required, 2 in the example
  P is the probability of failure of each item

Note that the first term is the number of combinations

$$(4 \times 3 \times 2 \times 1) / [2 \times 1(4-2)!] = (4 \times 3) / (2 \times 1) = 6$$

as derived above. However, in calculating failure combinations, it is important to be sure to identify all failure modes, bearing in mind that failure is random and is by chance. In the case of the delivery vans where there are four but only three are needed, the failure modes when three are not available will be:

1. 4 out of 4 failed *or*
2. 3 out of 4 failed *or*
3. 2 out of 4 failed

All these failure modes will be unacceptable and therefore the probability that they can occur must be calculated; because none are acceptable, they constitute a series system, as characterized by *or* logic, and the results of each failure mode must be added together.

8.5.5   *Example of a diesel engine powered ship*

In the 1950s, it was proposed that ships could be driven by multi-engine installations consisting of eight small engines instead of two large ones. The advantages claimed were twofold.

1. Being small and lightweight, they would be easy to remove and replace and so reduce time in port for maintenance.
2. Breakdown of one engine would leave seven running and therefore they would be more reliable.

Risk analysis shows that this is a series arrangement, as they must all be working to drive the ship. Two scenarios can be compared and evaluated for each of the two configurations.

Assumptions

Two identical ships to be compared, each installed with the same power output, one ship with two engines, and one with eight engines. Consider the ability to cruise with all engines running at 75 per cent power, and alternatively at reduced speed with only half the engines running (100 per cent redundancy).

- The large engines are more reliable, with $P = 0.1278$.
- The small engines are less reliable, with $P = 0.273$.

Two-engined ship A

Cruising with both engines throttled back at 75 per cent power:

Probability of only one engine failed out of two, A or B = 2P

Reduced speed, one engine running (100 per cent redundancy):

Probability of one engine not running (both fail), A and B = P^2

Eight-engined ship B

Cruising with all engines running throttled back:

Probability of one engine failed out of eight, A, B, C, D, E, F, G or H = 8P

Cruising with six engines out of eight (alternative strategy):

Failure modes are   8 out of 8 or
                      7 out of 8 or
                      6 out of 8 or
                      5 out of 8 or
                      4 out of 8 or
                      3 out of 8

When running at half-power with four engines out of eight (100 per cent redundancy), failure modes end at five out of eight.

The binomial distribution can be calculated using a spreadsheet. n is 8 and $n!$ is $8 \times 7 \times 6 \times 5 \times 4 \times 3 \times 2 \times 1$. P is 0.273 and $(1 - P)$ is 0.727.

Table 8.3 Binomial distribution probable failure

r	$n!/r!$	$(n-r)!$	$n!/[r!(n-r)!]$	P^r	$(1-P)^{(n-r)}$	P_{system}
8	1	1	1	0.000 03	1	0.000 03
7	8	1	8	0.000 113	0.727	0.000 657
6	56	2	28	0.000 414	0.5282	0.006 122
5	336	6	56	0.0015	0.3842	0.032 27
			The probability of failing to provide four engines			0.039 08
4	1680	24	70	0.005 554	0.2793	0.1085
3	6720	120	56	0.0203	0.2030	0.2308
			The probability of failing to provide six engines, sum of r 8 to 3			0.378 38

Comment

The results of the above calculations are shown entered into Table 8.4, which shows the comparison between the two ships. The probability of failure and the resultant reliability of the two different ship engine installations are given. Note that it is normal practice to first calculate the probability of failure and then to find the resultant reliability.

Table 8.4 Comparison of ship A with ship B

	P_{system}	*Reliability*
Ship A		
Probability of two not available	$2P = 0.2556$	0.744
Probability of one not available	$P^2 = 0.016\ 33$	0.984
Ship B		
Probability of eight running	$8P = 2.184$	Two engines always fail
Probability of six running	0.378 38	0.6216
Probability of four engines running	0.039 08	0.9692

It can be seen that the multi-engine ship is less reliable than the twin-engine ship. If all engines are in use, then in the eight-engined ship two engines need maintenance all the time. When running at 75 per cent power it is better to run with six engines than to run with eight throttled back. Even so, the reliability of the multi-engine ship is still significantly less. At half-power the difference between them is less. They both have 100 per cent redundancy, but the multi-engine ship is still less reliable.

At the time, a demonstration multi-engine ship was indeed built. In practice it was found to be unreliable and could not maintain the specified cruising speed. Engine failure was a constant problem. On one occasion it had to be towed back to port due to engine failure. Needless to say the idea had to be abandoned. If reliability analysis could have been carried out at the time, a lot of expense and trouble could have been saved.

8.6 Fault tree analysis (FTA)

This is called a top-down method as opposed to the FMEA, which is a bottom-up method of analysis. To start with, the undesired top event has to be identified. From this point, all the possible events that could cause the failure then have to be identified. The procedure is then repeated for each subevent and so on until all the basic bottom events have been reached. A diagram has to be constructed and a probability failure rate assigned to each event so that the probability for the final top event can be calculated. It will be found that the basic bottom events will normally be basic engineering components for which data are available from data books. In the examples that follow, the data will be taken from the table of generic failure data given in Chapter 7.

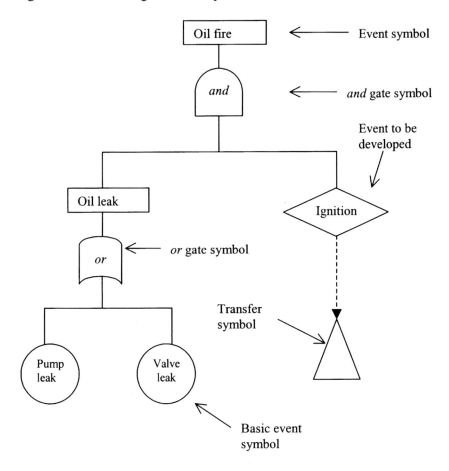

Fig. 8.5 Development of events leading to a fire

8.6.1   Fault tree development

The development of an FTA of a fire also serves to show the symbols used for fault trees. The transfer symbol allows other parts of the tree to be developed elsewhere.

8.7 FTA applied to the air pressure control system

In Chapter 5, the air pressure controls of an air starting system were studied using FMEA and HAZOP techniques in order to reduce their risk of failure. FTA can now be used to quantify the effects of the various design changes proposed. The fault tree can be developed as follows.

- *The top event:* this is over-pressure (explosion).
- *Second level events:* pressure relief valve fails (basic event) *and* compressor shut-down fails.
- *Third level events:* switchgear fails *or* pressure control fails.
- *Fourth level events:* manual control fails *and* auto-pressure control fails *and* high-pressure alarm/shut-down fails.
- *Manual control fails:* because operator fails *or* pressure gauge fails *or* pushbutton fails.

The drawing for the fault tree is shown in Fig. 8.6.

8.7.1   Common mode failure

The fault tree can also be constructed differently.

- *Second level events:* pressure relief valve fails (basic event) *and* pressure control fails.
- *Third level events:* manual control fails or auto-pressure control fails *or* high-pressure alarm/shut-down fails.
- *Manual control fails:* because operator fails *or* pressure gauge fails *or* pushbutton fails *or* switchgear fails.
- *Auto-pressure control fails:* because auto-pressure control fails *or* switchgear fails.
- *PAHH fails:* because high-pressure alarm/shut-down fails, *or* switchgear fails.

Here the same switchgear appears in three places; this is called a common mode failure. If not corrected, it will result in the failure of the switchgear being accounted for too many times.

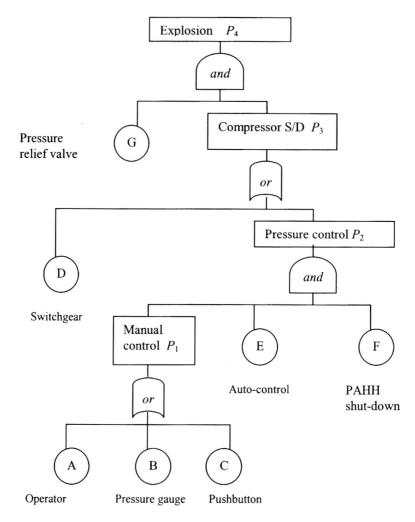

Fig. 8.6 FTA air pressure control system

Formulating

$P_1 = A + B + C$

$P_2 = (A + B + C)EF = AEF + BEF + CEF$

$P_3 = P_2 + D$, substituting for $P_2 = AEF + BEF + CEF + D$

$P_4 = (AEF + BEF + CEF + D)G$, probable risk of explosion

8.7.2 Boolean algebra

In a complex FTA it may not be easy to spot common mode failures and the use of Boolean algebra will correct for any common mode failures that may have slipped in. This is done by giving symbols to each element to enable algebraic expressions to be formulated for each failure event. The use of the Boolean algebra reduction laws will then eliminate any common mode failure errors from the final calculation.

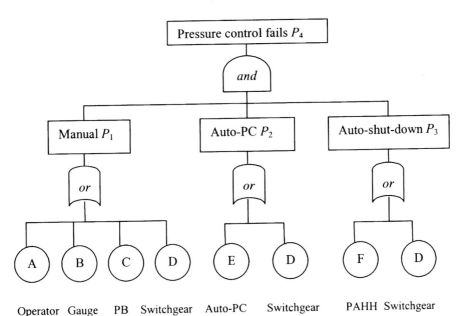

Operator Gauge   PB Switchgear Auto-PC   Switchgear   PAHH Switchgear

Fig. 8.7 FTA with common mode failure

Formulating

$P_1 = A + B + C + D$

$P_2 = E + D$

$P_3 = F + D$

$P_4 = P_1 \times P_2 \times P_3$ therefore

$P_4 = (A + B + C + D)(E + D)(F + D)$

Multiplying the first two brackets

$$P_4 = [AE + BE + CE + DE + (A + B + C + D)D](F + D)$$

Then rearranging

$$P_4 = [AE + BE + CE + (E + A + B + C)D + (D \times D)](F + D)$$

Applying Boolean reduction

$$(D \times D), (D + D), [D + D(A + \dots)], \text{ all} = D$$

$$P_4 = [AE + BE + CE + (E + A + B + C)D + D](F + D)$$

Reducing again

$$P_4 = (AE + BE + CE + D)(F + D)$$

and expanding and reducing again

$$P_4 = AEF + BEF + CEF + D$$

As found without common mode failure in the previous formulation.

8.7.3 Evaluation of results from the FTA

Table 8.5 Quantified risk of explosion

Item	Symbol	Gate	P/M	Evaluation
Operator	A	or	350	
Pressure gauge	B	or	33	$P_1 = A + B + C$
Pushbutton	C	or	0.8	$P_1 = 350 + 33 + 0.8$
Manual control	P_1		383.8	$= P_{2a}$ manual alone
Auto-pressure control	E	and	72	The most reliable
PAHH shut-down	F	and	4000	
Auto-PC + PAHH	P_{2b}		0.288	$P_{2b} = E \times F$
Manual + Auto-PC + PAHH	P_{2c}		0.001 09	$P_{2c} = P_{2a} \times E \times F$
Switchgear	D	or	480	
Compressor shut-down manual	P_{3a}		863.8	$P_{3a} = P_{2a} + D$
Compressor shut-down, auto-PC	P_{3b}		552	$P_{3b} = E + D$
Compressor shut-down, auto-PC + PAHH	P_{3c}		480.288	$P_{3c} = P_{2b} + D$
Compressor shut-down, manual + auto-PC + PAHH	P_{3d}		480.001	$P_{3d} = P_{2c} + D$
Pressure relief valve	G	and	65	
Explosion with P_{3a}	P_{4a}	$= P = 0.056\ 14 \times 10^{-6}$		$(P_{4a} = P_{3a} \times G)$
Explosion with P_{3b}	P_{4b}	$= P = 0.035\ 88 \times 10^{-6}$		$(P_{4b} = P_{3b} \times G)$
Explosion with P_{3c}	P_{4c}	$= P = 0.031\ 22 \times 10^{-6}$		$(P_{4c} = P_{3c} \times G)$

Equipment failure rates from Table 8.2 can be used so that the quantified risk can be obtained. The probabilities of failure for the different pressure control configurations are shown in the table, together with the resultant probability of an explosion.

8.7.4   Analysis of results

The probability of failure of the manual control is five times more than the use of an automatic pressure control (PC). The risk of failure with the addition of a PAHH (pressure alarm high, high) to the automatic control is further reduced. The risk is hardly affected when the manual control back-up is added. Note that this is obvious by inspection because P_{3d} hardly differs from P_{3c}. Manual control, however, will still be needed for emergencies and for test running after maintenance.

If reliability was to be the acceptance criterion, it will be noted that all the results are over 99 per cent reliable. It is the consequence of the control failure that is the determining factor, which is to avoid an explosion.

Examination of maintenance test intervals

The PAHH has a P value of $4000/10^6 = 0.004$ for $T = 1000$ h. If the test interval is increased by five times to $T = 5000$ h then the P value would be $20\,000/10^6$ or 0.02. The effect on the probability of failure is shown in Table 8.6.

Table 8.6 Effect of testing period

Auto-pressure control	E	and	72	
PAHH shut-down	F	and	20 000	With T of 5000 h
Auto-PC + PAHH	P_{2b}		1.44	$P_{2b} = E \times F$
Switchgear	D	or	480	
Compressor shut-down, auto PC + PAHH	P_{3c}		481.44	$P_{3b} = P_{2b} + D$
Pressure relief valve	G	and	65	
Explosion with P_{3b}	P_{4b}		0.031 29	$P_{4b} = P_{3c} \times G$
Explosion with P_{3b}	P_{4c}		0.031 22	With T of 1000 h

It can be seen that the probable failure of the PAHH does not seriously affect the chance of an explosion as compared with the results shown in Table 8.5. To understand the situation more fully, the concept of 'demand rate' is needed.

The automatic PC has a probable failure of $72/10^6$ h. That is 1/13 889 h. The PAHH, therefore, only probably needs to function once every 13 889 h. Although there is a temptation to further extend the testing interval, it is prudent to keep it below half the demand interval as a maximum.

In the same way, the testing interval of the pressure relief valve (PRV) can also be examined. The probability of an explosion is given by

$$P_{4b} = P_{3c} \times G = 0.031\ 29/10^6$$

The PRV has a probable failure rate of $65/10^6$ for a testing interval of 500 h. If the test interval were extended by 10 times to 5000 h then

$$P_{4b} = P_{3c} \times G = 0.3129/10^6$$

This is still well below $1/10^6$ and is still very safe. Note, however, that the test interval of the PRV, depending on its application, may well be dictated by the regulations appertaining to pressure systems, and this must be verified.

Significance of the PAHH
Examination of the figures show that the probability of failure of the automatic PC is 200 times greater than when there is a back-up PAHH. The calculations also show that the PAHH is only called upon to function once every few years. *This is a very important point.* To the operators, the PAHH is useless because it never does anything, and yet it has such significance on the pressure control system reliability. It has been recorded that in one plant there was just such a situation. The back-up device was causing spurious trips. The plant functioned quite well without it and so it was disconnected. There were no operating problems and it was forgotten about until a few years later, when the event that never happens, happened. There was no back-up. Disaster struck.

Cyclic operations
The analysis so far has been based on continuous operation. The air system, depending on the type of operation, could be operated for a short period of time for a number of times in a year. An air starting system for a diesel engine is used and then recharged, ready for the next start-up requirement. As an example, the case of an air starting system on a ferry ship can be considered. Demand rate is then the number of times it is needed per year. Hazard rate is the number of times it might fail.

- *Compressor shut-down demand rate D:* 300 times a year *or* 300/8760 h
- *Compressor shut-down P_{3c}:* $480/10^6$ h
- *Shut-down hazard rate H:* $(300/8760)(480/10^6) = 16.44/10^6$ h
  or once every 7 years

- *PRV demand rate D_2:* 16.44/10^6 h
- *PRV failure probability (G from table):* 65/10^6
- *PRV hazard rate:* $(65/10^6)(16.44/10^6) = 0.001\ 068/10^6$ h
  or once every 106 886 years

Conclusion

It has been shown in Table 8.5 how the pressure control system can be improved from a manual one to one with automatic PC + PAHH as follows:

Type of control	Probability of failure P	Reliability $R = 1 - P$
Manual	383.8 per 10^6	99.96%
Auto-PC + PAHH	0.288 per 10^6	99.999 97%

It has also been shown that the time to failure depends on how often the system is used. This can be further illustrated by considering a manufacturing operation of a 40 h week and a process plant operation that is continuous, 7 days a week throughout the year.

Type of operation		Breakdown	
h/week	h/year	Manual	Auto-PC + PAHH
40	2080	0.798/year	1/1668 years
168	8736	3.35/year	1/397 years

This clearly shows that for continuous operation the higher failure rate of a manual control system would be troublesome and that the use of the automatic PC + PAHH is more desirable.

It should also be noted that for a machine or plant to achieve, say, 97 per cent reliability (a failure rate of 3 per cent), and if there are ten subsections that must all work, then the failure rate for each has to be 0.3/10. That is 0.03. Each has to be 99.97 per cent reliable. This explains why the reliability of a control system of 99.999 97 per cent is not surprising.

It can be seen, moreover, that the possibility of the compressor not shutting down is dominated by the switchgear reliability. As would be expected, the risk of an explosion is very much affected by the pressure relief valve. The results do show that the risk of an explosion is slight, whatever the control system.

The analysis also allows study of the effects of the selected test intervals. This is important as it affects the maintenance costs, which must be balanced with safety.

Because the law requires all risks to be reduced 'so far as is reasonably practicable', use of a manual control system would not be acceptable. The cost of using automatic pressure controls with PAHH back-up is not excessive and would be considered both reasonable and practicable. For a system that is in cyclic demand the risk of an explosion is so low as to be non-existent.

8.8 Some other concepts

The basic concepts for risk analysis have been given but there are some other important concepts that must be considered in design.

Common mode failure
In the example of the delivery van, it was shown that having spare vans gives redundancy so that if one van failed, another was available to be used. In the event of a traffic jam the man would fail to deliver – spare vans would not help. This would also be the case if a flood made all roads impassable. This shows that redundancy does *not* provide reliability if there is a common failure mode.

Diversity
In the case of the man unable to deliver due to a traffic jam or floods, if he also had a helicopter, a bicycle or an amphibious vehicle he would have overcome his problem. This shows the principle of diversity as well as redundancy. He has more than one type of vehicle and more than one way of doing the job.

Fail-safe
This uses the idea that should anything fail, safety is not jeopardized; for example, the use of electrical switches that cut power when they fail. This is usually used for controls. Control valves can be arranged to fail in a safe position. It improves safety but reduces reliability.

Segregation
If all the delivery vans were parked at the forecourt of the warehouse, and a broken-down truck blocked the exit, again this would be a common mode failure. The *consequence* of the truck failure caused the problem. The problem could have been avoided by segregation, dispersing the vans to park at different locations.

Engineers should use these concepts to check their designs and plans in improving reliability and safety (bearing in mind that the one does not always ensure the other – they could be in conflict).

8.9 Risk assessment

In the example given of the risk analysis of a pressure control system and the possible risk of an explosion, the work is not complete until there has been a risk assessment. In order to meet the economic objectives of a plant or machine it is necessary to verify that it will meet its reliability targets. For the purposes of ensuring the health and safety of people, a risk assessment is required.

Risk assessment, besides the quantification of the risk, also needs an appraisal of the consequences. The following questions need to be answered:

1. Where is the hazard located?
2. What will be the consequential damage?
3. What is the risk from the consequential damage?
4. How many people could be in the vicinity?
5. Would the public be affected?
6. What injuries could be sustained?

Example: risk assessment of an air receiver explosion

It has been calculated that the risk of an explosion can be reduced to a probability of 0.031 22 per 10^6 h. Assuming continuous operation with an annual shut-down for maintenance, the hours per year is 8000 h.

The probability of an explosion (hazard rate) will be 1/4004 per annum. To assess the consequence, the explosion must coincide with the number of people present and the possible number killed must be calculated.

Location

The receiver is located in a compressor building. The building has one wall adjacent to a public road with a busy footway.

Consequences of an explosion

In the case of rupture, the air receiver is likely to split along its axis where it is most highly stressed. It is likely to be along the welded seam, which will be weaker than the parent metal. However, the effects of corrosion could produce more highly stressed areas and so the location of the rupture is uncertain. The direction of the pressure wave therefore cannot be predicted with certainty. Whatever the direction there are no items that could be damaged by the blast. Other contents of the room are compressors and motors and their associated pipework, all of which are securely bolted down. Electrical panels and control panels could be damaged but they are shielded from a direct line of sight to the air receiver. The blast is not contained as there are air vents and windows in the room and so the glass of the windows will be blown out.

The risk due to the consequential damage
The most serious risk will be due to the loss of utility air. As there is more than one receiver it is possible that only one has ruptured and so air supplies can be restored quickly. The plant is safeguarded by an emergency shut-down system. It is likely that damage to the building will be limited to the glass in the windows. The flying glass from the windows is in the direction of a public road that is in daily use with many people passing by. Other windows face into the plant, which is a bulk storage area.

Risk to workers
The compressor house is unmanned and there is an annual shut-down for maintenance. A team of six workers cover continuous operation with three shifts and a rota system. In an 8 h shift one person could be next to the air receiver for 10 min. The chance that a person could be exposed is

$$10/(8 \times 60) = 0.021 \text{ of the time}$$

For someone to be killed, they must be there *and* the explosion occurs. Therefore the chance of being there, *times*, the probability of an explosion gives the probability of a person being killed

$$0.021 \times 1/4004 = 0.5 \times 10^{-5} \text{ dead/annum}$$

As there are six workers the risk to one of them will be 1/6. For a basis of comparison it will be necessary to check the risk to 10^5 persons/annum:

$$(0.5 \times 10^{-5})(10^5)/6 = 0.0833 \text{ which is acceptable (see Table 8.1).}$$

Risk to the public
Any explosion will cause flying glass to injure members of the public. During football matches the pavement outside exposed to the windows could contain hundreds of people. This is where a bus stop is located. Normally being the route to the market, there could be tens of people here. Buses pass by frequently at 5-min intervals.

Conclusion
The possible risk to workers as a result of an explosion will be less than one in a 100 million. This is very safe and is acceptable. The risk to the public, however, is very high. If there is an average number of 20 people present in the event of an explosion, then the probability of people being

injured (assuming the same exposure time) will be 20 times the probability of injury to a worker. This is tolerable but needs justification.

In accordance with the preferred hierarchy of risk control, the risk to the public should be avoided if possible. Relocating the air receivers outside the compressor house, on the other side away from the road, can do this. The cost impact would be minimal. The danger to workers is unaffected, which in any event is much less than one in a million.

8.10 Summary

This chapter has shown how to deal with 'as low as reasonably practicable' and demonstrated the benchmarks against which risk can be judged. The availability of equipment failure data and the ability to use this for systems using FTA provide a powerful tool. This enables engineers to judge how best to optimize their designs for safety and reliability. It enables the application of redundancy and the effects on reliability to be evaluated. Furthermore, it provides guidance on the necessary maintenance actions needed to ensure that the design objectives are met.

It has also been shown that to find the possibility of a risk is not enough and that the consequence of the risk must be assessed to ensure safety.

8.11 References

(1)   HSE (1999/2000) *Safety statistics bulletin*, HMSO, London.
(2)   HSE (1992) *The tolerability of risk from nuclear power stations*, HSMO, London, ISBN 0118854917.
(3)   **Davidson, J.** (1988) *The reliability of mechanical systems*, IMechE, ISBN 0 85298 881 8.

Chapter 9

Risk Management

9.1 Introduction

The risks to a business are managed when all possible hazards have been identified and actions formulated to control the risks that could arise from them. Engineering is of course a business; it is a means of producing products and services for society at large. While engineering is an arm of management and the means by which products and services are produced, engineering alone cannot manage risk. The management of risk is a partnership between those who enable, conceive, produce and use the products and services. Each of their roles can be defined as follows.

1. Corporate management enables things to be done because it controls the finances, sets objectives and assigns responsibilities.
2. Designers conceive ideas and turn them into detailed drawings and specifications.
3. Engineers turn detailed drawings and specifications into plant and equipment.
4. Operators and users put plant and equipment to useful purpose.

As corporate management controls finance and sets objectives, the management of risk must start there. Risks, however, can only be managed if they are recognized as a threat and there is a fear of their consequences. Most disasters occur precisely because the risk was not recognized or feared. Unfortunately some boards of directors lack the imagination or experience to recognize the risk until disaster strikes. In some circumstances they are even unable to respond to warnings given and to heed the requests for action.

Engineers and operators, being closer to the events, are more able to recognize hazards and the possible risks. When disaster strikes, they too must bear some responsibility because of their inability to convince corporate management of the possible dangers and the need for action. They must be responsible for identifying hazards and must learn how to make a case for action that is persuasive to corporate management.

9.2 The cost of safety and reliability

Risk management costs money. A case has to be given to spend the money and to justify its use. The cost can be considered as an insurance premium paid to protect the loss of an asset. The money to be invested can then be evaluated against the benefit gained. However, it should be recognized that all loss is not directly financial. Loss of public confidence will ultimately result in loss of revenue, which can be measured, but the cost of regaining it will be more difficult to assess, see Table 9.1.

Table 9.1 Assets at risk

Risk to	Hazard	Financial loss	Other loss
Plant, facility	Fire, explosion, failure to produce output or quality	Capital, product cash flow	Customers
Output	Reliability	Cash flow	Goodwill
Quality	Failure of QC	Warranty claims	Goodwill
Reliability	Poor design	Cost of modifications	Goodwill, sales loss
Workers	Accidents and fatal injuries, effect on health due to emissions	Compensation	Lost time, loss of expertise, need to train new workers
Public	Ditto	Compensation	PR, goodwill, political repercussions

9.2.1 The true cost

Most companies have insurance cover for liabilities due to injuries and the ill-health of employees, for third-party claims, and for plant and buildings. Costs not covered by insurance can include:

- sick pay;
- damage or loss of product and raw materials;
- repairs to plant and equipment;
- overtime working and temporary labour;
- production delays;
- extra cost of temporary contracting out;
- renting of temporary premises;
- investigation time;
- fines.

Studies by the HSE have shown that uninsured losses in a year for a range of typical businesses could range from 2 to 36 times the premiums paid for insurance cover. On average, for every £1 paid in insurance premiums, £10 was spent on uninsured costs.

It is possible to determine the savings to be made by reducing the number of accidents, or avoiding them, over the lifetime of a plant. These savings, using accounting methods such as discounted cash flow, can then be converted to present-day value and compared with the cost of investment for safety. However, other cost factors may well be overriding.

Reliability, where safety is not an issue, is relatively straightforward. It affects cash flow because an unreliable plant or machine disrupts its utilization. Unfortunately this issue, although so obvious, is very often missed in the financing of projects. Reliability is of course included in the overall concept, which will be the basis for project financing. Due to the subsequent stringent cost control to complete on time and within budget, any extras outside of the project scope cannot be included. This is usually brushed aside on the basis that it should come out of the operating budget, as an operating improvement. Any add-on safety or reliability features after project completion will always be more expensive.

9.2.2   Other costs

These in most cases are even more compelling than any direct financial cost for safety and reliability. The general concern for safety and the unwillingness of the public to accept accidents and risks from the engineered infrastructure and services have resulted in legal requirements on management to manage risk.

The general public has also become more educated and aware of safety and reliability issues. As a result there is media attention on any shortcoming, ranging from the reliability of railway timetables to the reliability of buildings to withstand earthquake. Any fatal injuries caused by industrial disaster become a focus for debate. The public know their rights and are eager to seek redress through the courts of law. The following examples illustrate the effects of this.

- The Mercedes A model was found to be unstable when driven at extreme limits. It was withdrawn and modified. A decade earlier, the same criticism of a vehicle was levelled against another manufacturer. It was ignored with no repercussions and it was considered that the onus was on the driver to drive with due care.
- An oil company decided to decommission an off-shore oil rig and to sink it. Public outcry and the media spotlight on the environmental impact caused the company to change its plans.
- As a result of media attention and public outcry following a series of fatal accidents, the chief executive of the company running a railroad had to resign and the organization of the company had to be restructured to focus on the risk to safety in its operations.
- The collapse of a bridge in Portugal resulted in the resignation of a government minister.

9.3 Safety culture

It has been said that people cannot be changed and that therefore the work situations have to be changed. While it is possible to engineer systems that will reduce the risk of human error, experience has shown that this is not enough. People's attitudes can and must be changed. A safety culture has to be created, which is the purpose of the regulations on managing risk.

9.3.1 Safety policy

A safety policy is required in the UK by Section 2(3) of the Health and Safety at Work Act. The management has to draw up a safety policy statement that must be displayed and issued to all employees. It must also be reviewed and updated on a regular basis. It has to be in three parts, as follows.

General policy

This gives the general objectives of the company in the form of a mission statement. This must be all-embracing with regard to the expected attitude of all parties (management, supervisors and workers) in achieving the desired objectives, together with the common benefits to be gained.

Organization

This should identify the persons and their departments, as applicable, for supervising, enforcing and monitoring all required safety actions and the recording of any failure in compliance with any resulting injuries. The persons responsible for the following functions should be listed together as to how they can be contacted.

- reporting investigations and recording incidents;
- fire precautions, fire drill and evacuation;
- first aid;
- safety;
- training;
- statutory inspections and maintenance of safety equipment;
- management of (plant) change.

The arrangements for health and safety

This should list all the identified hazards to the workforce and to the general public, especially those that depend on the need to comply with working procedures. The hazards should be listed, together with the provisions to avoid them, showing those that involve people and what their duties are. The arrangements for reporting perceived hazards, dealing with any emergency and the arrangements for training, supervision and enforcement of safety procedures must be made clear.

It is important that all are involved and consulted on the arrangements so that there is a common ownership. Once the arrangements have been established it is then the management's task to control and enforce them by strict monitoring of performance. The measures required are explained below with some examples of control procedures that illustrate the general approach to be adopted.

9.4 Control

Active steps are needed to control safety and reliability. Things cannot be left to happen by chance and a number of important functions must be formally managed to ensure that their required objectives are met.

9.4.1 Quality control (QC)

Quality has no meaning unless there is a means of measuring it, with limits set as to what is acceptable and not acceptable. This principle can be applied to a whole range of activities from production processes to human behaviour. It is increasingly being applied on all types of operations as a means of measuring improvement and gains in efficiency.

9.4.2   Quality assurance (QA)

QC is the act of measuring quality. QA is the act of verifying that it has been carried out. Unless the actual measurement has been recorded and the act has been witnessed, QC really has no value. The statement 'of acceptable quality' has no meaning.

QA for safety procedures is essential. Records that instruments and safety devices, which are subject to failure on demand, have been tested at the prescribed intervals are essential. Everyone needs to understand this concept. Safety regulations rely on proof of innocence in the event of a disaster; if this cannot be proven then due care has not been exercised.

9.4.3   Permit to work systems, the control of hazardous energy, lock-out/tag-out, references (1) and (2)

This is a specific QC/QA system intended to ensure that work can be carried out safely on plant and equipment which normally present hazards to their access. The operation of a permit to work system will cover a number of phases. These must be treated separately, supervised and monitored.

Take out of service

While in service, plant and machinery will be under the control of the shift operators. It could be serving a critical function that, if disturbed, may pose a danger to the process. To avoid any misunderstanding, there must be a formal handover of jurisdiction and formal out-of-service notices will need to be posted.

Isolation and making safe from the plant

The equipment or plant must be isolated from the process and made safe. This means that any inventory must be safely discharged, the system must be purged to ensure that no contaminants remain, and then gas tested to prove that it is safe. The area will have to be cordoned off, out of bounds to anyone without a permit. All isolation valves and controls that are not to be operated must be labelled as such. Ideally they should also be locked or have blinds installed which cannot easily be removed.

Isolation and making safe from other feeds

The equipment or plant must be isolated from other feeds such as electrical supplies, and consumables such as chemicals, fuel, etc. Up to this point, the plant is usually still under the jurisdiction of the operating department.

Safe for access
After making safe, verified by inspection, a permit to work can then be issued. Handover of jurisdiction is given to the maintenance department. Workers are then allowed to enter to carry out the required work.

Work complete
On completion, the work has to be inspected, checked as acceptable and the permit to work withdrawn. Jurisdiction is formally handed over to the operations department.

Recommissioning
The equipment or plant has to be functionally checked and tested as being acceptable for returning to service. This is usually the responsibility of the operations department. The equipment or plant will need progressive reconnection and the removal of isolation tags in the process.

Handover
Shift plant operators need to formally take over jurisdiction, remove out-of-service notices, and prepare the equipment or plant for operational use.

9.4.4 Implementation
The control of the process requires formalized documentation. The design of the paperwork must reflect the steps in the work process. A form of work permit as suggested by the HSE is given in Table 9.2. This reflects some of the important areas of concern, which need to be monitored.

Table 9.2 A form of work permit

---

1 Permit title   *2 Permit no.*

                                                        Other related permits.................

3 Job location

4 Plant or equipment identification

5 Scope of work and any limitations to what is allowed

6a Hazards due to the process

6b Hazards arising from the work

7 Additional precautions needed	*Completed*	*Date*	*Time*	*Signed*

a)...

b)...

8 Protective equipment to be used	*Issued*	*Date*	*Time*	*By*

....

....

9 Authorization for the work to proceed	*Listed precautions (7)*	*Verified*

a)...Signed...............

b)...Signed.................

Authorization date...........permit duration........................ Signed

10 Acceptance to confirm understanding of work to be done, hazards involved and the precautions needed, and that all workers have been informed. Signed.............

11 Shift handover procedures satisfied to be confirmed by old and new shift leaders
Hazards understood and safety checks verified   Remaining scope of work recorded
Remaining time period.................Signed......................Signed..................

12 Hand-back	Certify work complete, ready to recommission	Signed	Date
	Accepted for recommissioning	Signed	Date
13 Cancellation	Certify recommissioned, ready for operation	Signed	Date

---

Although QC/QA and permit to work systems should prevent errors, they do not. The UK national authority, the HSE, who monitor all accidents, report that a third of all accidents in the chemical industry were maintenance related, most of them being caused by a lack of, or ineffective, permit to work systems. This underlines the need for these procedures to be compiled by people who have the required detailed knowledge and expertise to ensure they are correct in detail. The ways in which they can be misunderstood or circumvented need to be identified and prevented.

Related permits
Very often, when any plant or equipment is shut down, it also enables access to other equipment. For example, if a boiler is shut down, work could also be done on the boiler feed water pumps. A work permit must also control this work. The work on the boiler and the feed water pumps will need co-ordination. The issue of all work permits must, therefore, be co-ordinated by a central office to ensure any inter-reaction is identified and controlled safely.

Identification
Just as the wrong identification of a patient in a hospital has resulted in the patient having the wrong organ removed, wrong identification in industry has similar results. Opening the wrong valve when men are working can cause fatal injury. Identification of the correct work area is critical to safety.

Scope of work
This should be strictly controlled. In a boiler inspection it is possible that some defects are revealed which need repair. This will extend the time needed for the shut-down; other work such as welding and radiographic examination may be needed, with new hazards to be considered. Any extension of scope needs to be controlled.

Hazards and precautions needed
These must be formally stated and emphasized to ensure that complacency is overcome. Management should strictly enforce this, as complacency could also extend to the supervisors.

Shift change
This is a situation of high risk due to misunderstandings that can arise. It is important for management to allow for adequate overlap of shifts to enable a formal handover, with a proper record of the progress of the work and the work yet to be completed.

9.4.5   Permit-required confined spaces, references (3) and (4)

A confined space is any location where entry, working space and exit is hindered. Any of the following hazards will establish a permit-required confined space:

- entrapment;
- engulfment by materials that are present;
- hazardous atmospheric conditions;
- confinement, has cramp, gets stuck, gets claustrophobia;
- restricted entry and exit;
- restricted airflow, could faint.

A written procedure with controlled access and documentation similar to that for controlling hazardous energy (see Section 9.4.3) is required. Furthermore the following extra, specific measures are required:

- entry must be barred to prevent unauthorized entry;
- a dedicated attendant must be stationed outside;
- testing of the atmosphere inside must be carried out before entry and during work inside to ensure that it is safe;
- the duty of those entering to maintain continuous communication with those outside;
- a trained rescue team must be on standby;
- the attendant must remain outside and summon the rescue team if help is needed;
- a supervisor must check that all procedures are carried out and that the entry permit is checked and verified. He must ensure that the outside attendant and those entering know their duties, and that a qualified rescue team is on standby.

The principles of the notification of hazards, instruction, training, recording and monitoring will apply.

9.5 Supervision

Successful risk management depends on dedicated supervision. This provides leadership, motivation and the co-operation of the workforce. It is important that there are at least two tiers: direct supervision and management.

The task of management is to ensure that procedures are rigorously tested before their introduction and that adequate training in their implementation is provided, with continuous monitoring to maintain standards.

9.5.1 Matrix management

It has been said that man (or woman) cannot serve two masters. In the case of risk management, this is especially true. There is a conflict of interest. Output versus quality. Performance versus safety.

Matrix organizations can resolve this problem. Individuals are then subjected to a tug of war, with them in the centre. It is the dynamic tension that ensures equal attention to opposing objectives.

If an operator who for years and years has aimed for production targets, of whatever sort, is suddenly posed with a situation that requires him to stop production, he becomes confused. If, week in week out, there are meetings to review production and he is praised for all his work, how can he suddenly without warning be expected to act out of character? To counteract this situation, there has to be a different organization involved which focuses on safety and makes equally strident demands.

9.6 Education and training

9.6.1 Definitions

Training means to bring to a standard of efficiency by a course of instruction and practice. To educate, on the other hand, means a course of instruction to develop mental powers. The two are different and have different objectives.

9.6.2 Education and training

Control room operators need to be educated to observe signals and instrument readings so as to form the correct conclusion, in order to decide a course of action. They have to be mentally conditioned to react in a certain way. The actions required of them are not demanding; they just have to know which control to adjust or what button to press.

Technicians, on the other hand, need to be trained to follow procedures and to carry out actions in a certain sequence. They need to be educated to understand the reasons for what is required so that they can decide what to do if things go wrong. They also need training so that they become accustomed to carrying out routines in a set way.

It is a common error to muddle up the difference between training and education and to know what is required in a given situation. Educationalists who condemned learning by rote and stated that true education was development of mental capacity have added to this confusion. The confusion was caused by the attitude that this was correct for all situations. When applied to the teaching of mathematics, for example, a whole generation of students were handicapped. They understood the theories of how to carry out algebraic manipulation but were unable to do so with speed and precision, which needs training. This handicapped them in learning more advanced mathematics where algebraic manipulation is taken for granted.

9.6.3 Motivation

Education is also useless if the effect is not measured. A form of examination is essential. This can be a written or oral examination and some form of incentive may be needed to motivate the employees to learn. Many industrial concerns give certificates and have some form of grading and presentation ceremony to underline the importance of the instruction. These are all measures by management to underline the importance of the instruction and increase the safety consciousness of the employees.

9.6.4 Practice and testing

Training can only be complete if a course of practice follows instruction. With work procedures, this practice can be by example, under supervision. This can be said to be on-the-job learning. Events unfold in an orderly fashion and can be observed and absorbed. The ability of the trainee can be verified by examination. The trainee can also demonstrate his ability to perform procedures.

In emergency situations, speed is of the essence. The person may be well educated and intelligent, but to rely only on their brain to observe, deduce and recall is too risky. Very often the emergency in question may never happen and, should it happen, it would be unexpected and without warning. Training simulators are used to overcome this problem. Airline pilots sit in a flight simulator cockpit and the surroundings appear as if they are flying in an aircraft. The instructor is able to feed in the signals for any type of emergency and the pilot must respond as if in a real aircraft. If necessary, they can be drilled until they have a perfect response.

In the Norwegian sector of the North Sea full-scale manoeuvres have been carried out. A control room monitor can be linked with a training simulator and an emergency situation developed to the extent of all emergency services being involved up to flying rescue operations.

In most industrial plant, these extremes may not be warranted, but the management will need to devise appropriate ways of testing operator response, even down to actually lifting a telephone to alert emergency services – in a panic would they remember the number to dial? To minimize risk, staff must be trained to deal with rare events and these exercises must be repeated regularly to be effective.

9.7 Monitoring performance

In the UK, in accordance with the Reporting of Injuries, Diseases and Dangerous Occurrences Regulations (RIDDOR), each plant is required to report and keep records of all industrial injuries and near-misses.

When an industrial injury or near-injury has occurred, it is also a measurement of failure. Other measurements that monitor performance are needed, such as:

- completion of training schedules;
- employee training achievements;
- verification of training quality and results;
- outcome of training exercises;
- schedule of spot checks and audits and their results;
- regular feedback meetings with all employees, with records of attendance;
- accident investigation and analysis (also required for near-misses).

The duty of management is to carry out on-the-job, random checks. This will verify that the requirements of the work permits and QC/QA procedures are being followed. In addition, a random selection of work permit records should be picked for formal audit. A check can then be made that the quality of records are acceptable and that no problems of application can be found by friendly cross-examination of the people involved.

It has been recorded that many accidents have taken place due to lax supervision. An experienced worker carries out a task for many years without incident and the supervisor has learnt to trust him. When the supervisor has other tasks to perform and he is busy, there is a temptation to sign off the check sheet without actually carrying out a personal check. This illustrates the need for dedicated supervision that remains focused and without distractions. The job of a management audit is to verify that short cuts do not take place and that supervision remains focused.

For risk management to be effective, performance must be measured, recorded and publicized. Good and bad practices when found should be made public for everyone's education, and any risk to safety shown to be against the common interest.

9.8 Emergency planning and management

In spite of doing everything possible, there will always be potential for events that lead to disaster. Adequate emergency planning will help to minimize the effects of the disaster. This is a legal requirement in accordance with the EC directive on Control of Major Accident Hazards (COMAH) and the Control of Industrial Major Accident Hazards Regulations (CIMAH) in the UK. In accordance with the COMAH directive, planning should recognize the need for the management of two areas: off-site, to deal with the external logistics, PR and the public; and on-site, to deal with the effects of the disaster. The OSHA has similar requirements **(5)**.

9.8.1   Off-site plans

These plans should include the following.

- Nomination of those who are authorized to set emergency plans in operation and who is to be in charge.
- Communications for early warning, alert and call-out.
- Arrangements to ensure the security of communications, with back-up provisions as necessary.
- Arrangements for co-ordinating and calling out external off-site emergency services.
- Arrangements for call-out of external emergency services for on-site assistance.
- Arrangements for dealing with the public and the media.

9.8.2   On-site plans

These plans should include the following.

- Those who are authorized to set emergency plans in operation and who is to be in charge.
- Person who is to liaise with the authority responsible for the external emergency plan.
- How to communicate with the authority responsible for the external plan. To alert the external authority, to set in motion the plan, the type of initial information required and the arrangements for the updating of information as it becomes available.
- Types of accidents that could occur and how they should be dealt with. What on-site resources are available and how they are to be deployed.

- The emergency procedures and the evacuation arrangements to be used.
- Training arrangements for on-site staff and co-ordinated exercises with off-site control centre and external services.

The speed of response to any incident will have an effect on its containment. This is the objective of preplanning and training exercises. Studies have shown that it is what happens within the first moments of an incident that will make the difference between a minor incident and a disaster. Adequate training and drill prevents escalation. This is illustrated by Fig. 9.1, see reference (**6**).

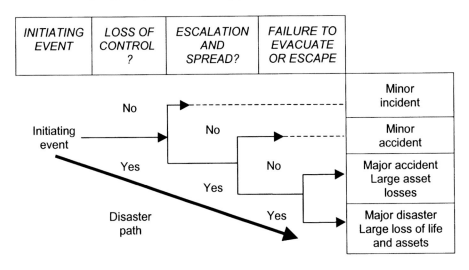

Fig. 9.1 The results of bad risk management

This simplified event tree illustrates the fact that matters escalate very quickly unless events come under control. Probably the disaster manager only has a few minutes to decide what to do in order to prevent escalation.

9.8.3 An imaginary disaster scenario

The incident
In most cases there is an alarm of impending danger and routine emergency shut-down (ESD) procedures take place. On the day in question there was no alarm, just a loud explosion. The noise was sufficiently alarming to cause the disaster management team to arrive at the plant

manager's conference room, which was the nominated disaster control centre (DCC).

The plant manager was already looking at the plant monitor to check the data logger and was speaking on the telephone to the shift supervisor, just as the operations manager was arriving in the plant control room. The explosion was reported to be in the boiler area and black smoke and flames could be seen. In the control room all the alarms were ringing. Some process units were in trouble due to depleting steam supplies. Furnace alarms were ringing as fuel supply pressure was falling.

The disaster manager

The plant manager, being the designated disaster manager, had the duty to minimize casualties, prevent escalation outside the works to safeguard the public, and to save the company's assets. He immediately instructed the plant manager to ensure that the plant was shut down safely and that all fuel systems were isolated. He then concentrated his mind on trying to foresee what might happen and to think of what to do to prevent it. The first thing of course was to sound the alarm for evacuation.

The incident co-ordinator

The safety manager, who was the incident co-ordinator, was the next to arrive at the DCC. He manned the multi-channel communication system and was in direct contact with the works first aid and fire fighting teams. Reports were coming in on the exact location of the incident, the extent of the fire and the first estimate of casualties. He immediately contacted the nominated local authority incident room. In the UK this is the Police, who in turn inform all the other services, who operate under independent command. In France it will be the Préfet or his nominated Sous-Préfet who has overall command. Other countries may have different arrangements. However, in the end the effectiveness depends on good co-ordination of all services involved.

He will also need to inform the gatehouse of the pending arrival of the external services and where they are to be directed. The works services must also be told of their arrival.

Incident log keeper

The plant manager's secretary, who had this role, was already at the DCC. Her first job was to check that the members of the DCC team were on their way. As the reports came in, she downloaded or entered them into the DCC computer. This had the facility to project the information on to a large screen on the wall of the DCC. The type of information to be logged had been agreed during training and she knew what to do. She also ensured that the tape-recorder was on so as to record all incoming and outgoing messages. This would be useful for the subsequent incident investigation.

Public relations

This was the human relations (HR) manager's job. She checked to make sure that the main switchboard was alerted and that all calls concerning the incident were directed to her dedicated telephone in the DCC. Her job was to prepare a press release and to produce a list of all employees on the site. The results of the shift team leader's roll call would come to her and this would show who was missing. The casualty list will also come to her so that all persons can be accounted for. The families of all those injured or missing will need to be informed and their questions answered. She will also need to keep check of casualties and their progress when they get to hospital.

Event log (dates and times to be recorded for each event)

1. The first inkling of the disaster.
2. The arrival of all members of the DCC team.
3. Arrival of the first information giving location and extent of the incident; external services informed.
4. Sound plant-wide alert.
5. Arrival of works fire fighters on the scene; and their report.
6. Issue public address to evacuate areas 6 and 7 (at risk) and the site of the incident, utilities area 5.
7. Dispatch safety officer to search area 6 and 7 for any stragglers.
8. Details of the incident: a boiler explosion with destruction of the boiler local control room and a fire caused by the fracture of fuel pipes.
9. External services informed and confirmation received that the fire and ambulance services had been dispatched.
10. First information of casualties.
11. External services informed of the number of casualties to ensure that the number of ambulances dispatched was sufficient.
12. Instructions issued to the plant control room to ensure the ESD of the fuel systems and to ensure that this extended to plot limits and plant battery limits. All the ESD valves were of the fire-proof type. Verification to be made of those nearest to the fire, at a safe distance, just in case they had been disabled. Alternative valves closed as necessary.

13. Confirmation received from the fire fighters that the fire from the fuel pipes had stopped. Burning fuel on the ground was spreading the fire and some nearby tanks were being engulfed.
14. Muster roll-call reports received. The people missing established, and information passed to the first aid rescue team.
15. Updated information passed to external services. Arrival of external services reported from the gatehouse and passed on to the works service teams.
16. Information received on the names of the designated hospitals, as more than one was needed due to the number of beds needed.
17. Missing persons found and accounted for.
18. First casualties leaving the works, identities established, initial assessment made and hospital destination known.
19. Families are informed and first press release issued.
20. Fire under control, no further spread.
21. PA announces incident under control. Alert cleared.
22. Incident contained within the works, no danger to the public.
23. Damage assessment and recovery planning instigated.

Tabletop exercises

Review of the above scenario demonstrates that disaster control is a matter of co-ordination and communication. The first instinct of the plant shift supervisor is to think of how to maintain production and get the plant under control. That has been their day-in and day-out job for years and so they will be confused as to what is the priority.

The job of the DCC team is to think out the priorities of what to do. How the incident was caused is relatively unimportant for the moment; only sufficient detail is needed to forecast its likely progression.

For the above reasons, the team can be trained by annual or biannual exercises around a table. A set incident routine can be followed so that each member can be exercised in his or her role. On other occasions, the exercise can be conducted jointly with the external incident team; it will be good for them to know each other anyway. Another important exercise is to test the communications systems. Precautions will need to be taken to cope with an incident in case this could occur while the exercise is going on!

9.9 Summary

The hazards in industry and the probable risks and consequences are many and varied. Whatever the situation, some form of risk management will be needed. The amount of effort needed will of course also vary, depending on the risk and its consequences.

The material provided in this chapter is based on the requirements for process plant. The basic principles are universal and so they can be adapted to control and manage risk for any given situation.

9.10 References

(1) HSE leaflet INDG98 rev3. *Permit to work systems.*
(2) OSHA 3120 *Control of hazardous energy, lockout/tagout.*
(3) HSE leaflet INDG258 *Safe work in confined spaces.*
(4) OSHA 3138 *Permit required confined spaces.*
(5) OSHA 3088 *How to prepare for work place emergencies.*
(6) **Strutt, J.E.** and **Lakey, J.R.A.** (1995) Education, training, and research in emergency planning and management, *Emergency Planning and Management*, IMechE Conference, Paper C507/009/95, Professional Engineering Publishing, ISBN 0 85298 854 7.

Chapter 10

Maintenance Strategies

10.1 Introduction

Plant and equipment, however well designed, will not remain safe or reliable unless it is maintained. In Chapter 6 it has been demonstrated how maintenance affects reliability and how items vital to safety may have unknowingly failed unless regularly tested.

Chapter 7 has demonstrated by statistical analysis that it is impossible to predict for certain when things will fail. The challenge for maintenance engineering is to make use of this knowledge and to achieve the best possible reliability and safety at the lowest possible cost. This has led to the development of a number of different strategies, and maintenance engineering has become a technology in its own right.

10.2 Breakdown maintenance

This strategy evolved from the early days of the industrial revolution and continued for many decades as a normal approach to maintenance. Even in modern times it is still widespread, with the homily 'if it ain't broke don't touch it'. Not a bad sentiment.

It fell into disrepute because no one could predict when things would break down. It was most inconvenient. In the old days when the operator was the engineer and the mechanic, the machine or plant just got nursed along.

With modern management, cash flow and reliable output are the watchwords, with a constant demand for improved output at lower cost. Breakdown maintenance became old fashioned and obsolete.

With modern developments and reliability analysis it has been realized that breakdown maintenance still has its place. In situations where it is not critical to safety and reliability, it still represents the lowest maintenance cost. It will also apply where there is adequate redundancy in a system and the increased risk during the downtime of one item is still acceptable.

10.3 Planned (preventative) maintenance

This was the first idea to evolve from breakdown maintenance. To avoid the disruption from random breakdowns, it was thought that maintenance could take place at planned intervals to ensure reliable operation.

This system is expensive because of the unpredictability of failure. Even to use planned maintenance to replace items that wear out (see Chapter 6) is wasteful and unreliable. It depends on the wear-out characteristic and the scatter of the values found. If the standard deviation is large, then as previously discussed, to ensure reliable operation, most of the items will be thrown away when there is still a lot of useful life left.

Equipment such as gas turbines are subject to planned maintenance. In these situations, wearing items such as combustion cans, nozzles and blades are inspected at planned intervals to check if they need replacement or repair. Typical planned intervals are:

- First hot path inspection: 12 000 h.
- Subsequent hot path inspection: 24 000 h.
- Major inspection: 48 000 h.

Major equipment manufacturers know the importance of reliability and the need for reduced downtime and maintenance cost. As operating experience is built up, they collect data and are able to formulate planned maintenance intervals. This will be based on how machines are used, the number of starts per year and other operating factors that will have an influence on component life.

10.4   Opportunity maintenance

Other names are also used, such as convenience or shadow maintenance. Large items that are subject to planned maintenance, or events that cause a shut-down, provide the opportunity to carry out other maintenance tasks. These can be carried out within the forced shut-down period so that another shut-down later can be avoided. This aids plant output and improves the overall plant availability.

As soon as the shut-down is known to be required, the necessary downtime has to be established. All other maintenance actions that are pending are then listed. Those that can be accomplished within the shut-down period can then be reviewed as possible candidates for maintenance action. It may even be of benefit to extend the shut-down to fit in more work.

When a planned maintenance operation is planned, the opportunity is known in advance and it is easy to plan other work, and this is usually done. In the situation of a forced outage, a rapid assessment is needed, first to estimate the expected downtime and then to adjust it when inspection reveals the full scope of work needed. Therefore, there are two phases in the planning of opportunity maintenance. A good maintenance database on computer is needed. This will enable speedy decisions to be made so as to take full opportunity of the time available. The possible extra work can then easily be selected and matched with the available resources.

10.5   Condition-based maintenance

This is based on the idea that it should be possible to measure the condition of equipment and so monitor its deterioration. This information then enables shut-down requirements to be predicted and planned. This procedure, however, is dependent on the ability to identify the parameters to be measured and the availability of instruments to perform the measurements. The following examples show the most common applications in use.

10.5.1 Vibration monitoring

It is well established that rotating machines exhibit signs of distress by how they vibrate. The various modes of vibration will indicate different defects. The information obtained and the method of analysis used are dependent on the type of instrument installed. These are listed below.

Non-contacting vibration and axial position probes

These are used for rotors or shafts running with electromagnetic or oil-lubricated sleeve bearings. Any rotor dynamic forces generated are counterbalanced by electromagnetic forces or by hydrodynamic forces, depending on the type of bearing used. Variations in radial forces will cause relative movement between the shaft and its bearings. Rotors for machines are usually also subject to axial forces and are restrained by thrust bearings. Oil-lubricated thrust bearings generate hydrodynamic forces to counteract any thrust loads. Variations in thrust load will cause relative axial motion between the shaft and the casing.

These instruments are used to measure the relative motion of the machine shaft and the machine casing. They respond to a change in the air gap between the probe and the shaft. The casing has to remain unaffected by the shaft movement. This is usually when the casing mass is very much greater than the rotor mass, as with heavy industrial machinery. Aero-engines, for example, use lightweight casings and must use a different type of instrument.

It should be noted that there are two types of rotor and bearing configurations in use. One is where the rotor weight is supported between bearings and the other is where the rotor is overhung and the two bearings are used to restrain the rotor overturning moment.

Two *X Y* radial probes are usually installed adjacent to each bearing, together with a phase measuring probe which monitors a mark on the shaft. The phase signal is related to the *X Y* probe signals, which then allows the relative direction of the shaft vibration to be known at each bearing location. Axial probes, mounted in the casing, monitor the movement of the shaft end. Table 10.1 gives some common failures and their signal characteristics. It is important to have a record of the machine characteristics in good running condition after commissioning for comparison, as the condition of the machine deteriorates. This is usually called the vibration signature. A change in the machine condition will cause a change in the signature. Software is available for this purpose.

Table 10.1 Common rotor failure characteristics

Failure mode	Cause	Signal
Unbalance	1. Residual unbalance 2. Uneven corrosion/erosion 3. Uneven deposits	Between bearing rotor Vibration at each bearing, in phase at running r/min Overhung rotor Vibration at each bearing 180° out of phase at running r/min
Subsynchronous excitation	Operating too close to the natural frequency Hydro/aerodynamic excitation	Increased vibration at frequencies other than running r/min
Oil whirl	Excessive wear/bearing clearance	Radial vibration at half-running r/min
Misalignment	1. Uneven settlement between machines 2. Bent shaft 3. Incorrectly seated bearings	High radial vibration at half-running r/min with high axial vibration. If measured at both ends of the shaft the axial vibration will be 180° out of phase
Tooth defect	Tooth damage in gearboxes	Increase in vibration at natural frequency and at tooth-passing frequency (No. of teeth × r/min)

Apart from obtaining the vibration signature and trending the deviation, it will also be necessary to compare signals with a recognized norm to judge when vibration is too high. Most manufacturers provide guidelines for this but there are also internationally recognized standards available, as shown later.

Accelerometers

These are solid-state devices with no moving parts. They respond to a wide band of frequencies and produce an electrical signal that is proportional to the vibration acceleration. They are widely used for machines where rotor vibration is sensed on the casing. This occurs where lightweight casings are used and/or where antifriction bearings are used.

They can either be hand-held for operator patrol measurement or permanently installed for continuous data collection. Different models tuned for a range of frequencies to suit different applications are available.

- Low frequencies to monitor cooling fans.
- High frequency, able to monitor blade-passing frequencies for gas and steam turbines.

Ball and roller antifriction bearings are used for a wide range of small turbomachines and electric motors, and they are the main cause of failure of these machines. The failure modes are cage wear/failure, ball damage and race damage. Cage failure is signalled by an increase in half-r/min vibration and ball or trace defects by harmonics of running r/min. Defect detection is complex and depends on the bearing details, such as ball diameter, number of balls and pitch diameter.

Accelerometers with software have been developed for the detection of antifriction bearing wear and fatigue. They use the Kurtosis technique for damage detection; further information can be obtained from detector manufacturers.

Velocity pick-up
These work by sensing the rate of change of flux in a sensing coil. Due to the use of moving parts they are less reliable than solid-state sensors. They are useful for monitoring machines with high levels of vibration at very high frequencies.

Vibration acceptance criteria
Internationally recognized acceptance criteria for factory testing of new machines as specified by the American Petroleum Institute (API) are given in Table 10.2. Manufacturers can also provide recommended alarm settings. They will need to be adjusted, based on operating experience.

Table 10.2 Vibration criteria

Machines with antifriction bearings	Type sensor	Location	API acceptance criteria
Centrifugal pump (note 1)	Accelerometer	Bearing housing	7.8 mm/s or 63 µm, whichever is less 5.1 mm/s filtered
General-purpose steam turbine	Ditto	Ditto	3.8 mm/s unfiltered 2.5 mm/s filtered

Machines with oil-lubricated sleeve bearings			Note: Nmc means maximum continuous r/min.
Centrifugal pump	Non-contact probe	Adjacent to bearings	10.2 mm/s or 63 µm, whichever is less 7.6 mm/s filtered
General-purpose steam turbine	Ditto	Ditto	$1.25 (12\,000/Nmc)^{0.5}$ mils or 50.8 µm plus run-out, whichever is less (note 5)
Special-purpose steam turbine	Ditto	Ditto	Ditto
Industrial gas turbine	Ditto	Ditto	Ditto
Centrifugal compressor	Ditto	Ditto	Ditto
Package integrally geared centrifugal compressors	Ditto	Ditto	Ditto
Special-purpose gearbox	Ditto	Ditto	$(12\,000/Nmc)^{0.5}$ mils or 50.8 µm plus run-out, whichever is less
Positive displacement screw compressor	Ditto	Ditto	$(12\,000/Nmc)^{0.5}$ mils or 63.5 µm plus run-out, whichever is less

Notes

1. These criteria are acceptance criteria on the test bed.
2. Velocity criteria are capped for low speeds on pumps and are limited by a maximum allowed peak-to-peak reading.
3. Pumps and general-purpose steam turbines fitted with antifriction bearings will generally suffer higher vibrations due to contributions

from harmonics. This is the reason why a lower reading is specified for measurements that filter out the harmonics.

4. The vibration measurement, in mils or μm, is peak to peak, or the double amplitude of vibration.
5. A mil is 0.001 inch or 25.4 μm. 1 μm is 0.001 mm.
6. Displacement or amplitude of vibration, α, when filtered for frequency is assumed to be sinusoidal. The following relationships are useful for conversion:

$$\text{Velocity} = 2\pi\alpha\ \text{Hz} \qquad \text{Acceleration} = \alpha(2\pi\ \text{Hz})^2$$

Alarm setting for maintenance

Premature maintenance is costly. Operators will therefore need to build upon their own experience for each machine and determine the level of vibration that needs action. For this to be done, the recording of baseline vibration signatures for each machine is paramount. Monitoring of trends on a specific machine basis will enable judgement on the machine's condition. Experience from a few shut-downs will enable adjustments to be made. It will be found that some machines are more sensitive than others to conditions that will cause excitation. One important criterion is the relative flexibility of the rotor. A sensitive rotor is one where the operating r/min:first stiff bearing critical speed is greater than 1. The gas density handled by a compressor is another. High gas density will result in more aerodynamic forces being generated. A combination of a sensitive rotor and high gas density can give rise to excitation at frequencies lower than the running speed. This is referred to as subsynchronous vibration. Centrifugal pumps, because they pump liquids, also experience these problems. To avoid these problems, stiff shaft rotors, with their first critical speed above running speed, are favoured. As a guide, a 12 mm/s velocity unfiltered reading should give cause for action unless experience proves otherwise.

Spectrum analysis

To enable vibration signatures to be obtained, real-time data capture with software for spectrum analysis is available. Some machines will exhibit vibration signals that are complex, due to the many forcing frequencies that may exist. This is especially true of pumps handling liquids, and compressors handling very high-density gases. They experience significant hydro- and aerodynamic forces. These tendencies are affected by the condition of wear rings, labyrinth seals and other changes in the fluid passages. For these reasons, spectrum analysis becomes important as it enables changes in condition to be more easily identified.

10.5.2 Efficiency monitoring

In a way, this can be more effective than vibration monitoring. Loss of efficiency is affected by wear, which can take place before hydro- or aerodynamic effects increase vibration. For static equipment, it may be the only way to measure condition.

Centrifugal pumps

For any given operating condition, any loss of efficiency will result in an increase in differential temperature across the machine. These differences will be small and the effectiveness of this procedure will depend on instrument accuracy. Specialist temperature measuring devices, developed for the purpose, are available. For certain situations, this is a very useful procedure.

Centrifugal compressors

As with pumps, for any given operating condition, any loss of efficiency will result in an increase in differential temperature across the machine. The temperature difference, more usually given as the ratio, is also affected by the gas composition, the volume flow and the pressure ratio. A sensitivity check will be needed to verify which parameters must be monitored, if not all of them.

Gas turbines

These are usually supplied complete with control panels which have data processing capability. Condition monitoring of the gas turbine compressor is usually standard, to indicate the need for compressor washing. Options for performance monitoring are available which will indicate deterioration of the hot gas path components.

Reciprocating compressors

These suffer from ring wear and valve deterioration. This results in the loss of volumetric efficiency. In multi-stage compressors, the preceding stages will have to work harder. The symptom is an increase in the preceding stage compression ratio and a loss of compression ratio in the affected stage.

Steam turbines

Steam turbines can suffer from the effects of poor steam quality that will result in blade deposits and steam path erosion. In the case of back-pressure turbines, the effect is shown by increased steam rate and reduced temperature difference. The monitoring of exit temperature may well be sufficient indication. In the case of multi-stage turbines, erosion and deposits will affect the first stages. An increase in initial stage pressure ratio will indicate deposits due to a reduction in area, and a reduction could indicate an increase due to erosion. The manufacturer should be able to advise on this.

Reciprocating internal combustion engines

Monitoring of the exhaust gas temperature from each cylinder provides an indication of combustion efficiency. Marine diesel engines usually include these in their standard scope of supply.

Lubricating oil

The efficiency of lubrication of machines depends mostly on the properties of the lubricating oil. Major capital equipment such as centrifugal compressors can have recommended planned maintenance intervals of 24 000 h. It has been reported that monitoring and maintaining the lubricating oil properties have enabled maintenance intervals to be extended significantly.

Heat exchangers

Heat exchangers will deteriorate in service due to deposits on the surfaces of the tubes or other heat exchange surfaces. There will be a loss of heat exchanged and operators will compensate for this by adjusting the flow. In time the exchanger will need to be cleaned. The MTTF for the exchanger will be known from experience. As the only thing that changes is the effective surface area, the log mean temperature difference (LMTD) has to change for the same heat duty. If needed, the monitoring of the LMTD will provide an indication of the condition of the heat exchanger surface area.

10.5.3 Other monitoring techniques

There are many other important methods to monitor condition, which will need to be considered by the maintenance engineer in the battle to control cost.

Perforation damage monitoring

On many plants, the use of seawater as a cooling medium is convenient, but leads to corrosion problems with a high maintenance cost. This is due to the need to re-tube a heat exchanger and to repair the effects of polluting the process stream.

1. Water-cooled gas heat exchangers are usually designed with the gas side at a higher pressure. The condition can be checked without internal inspection by isolating the waterside. Any high-pressure gas leaking into the waterside can by verified by the use of a gas detector at a high-point vent.
2. Seawater-cooled steam condensers suffer from seawater contamination of the condensate return, should there be a leak. Conductivity meters will detect contamination of the condensate.

Insulation condition monitoring
This can be monitored by the use of infrared imaging devices. These are very useful in finding, for example, furnace and boiler refractory damage while still in operation. This helps to plan for maintenance shut-downs in advance of internal inspection.

Acoustic monitoring
Fluid turbulence and leaks give rise to acoustic emissions and can be used to detect any abnormality. Systems have been developed to monitor pumps, transmission pipelines and mechanical seals. The problem has always been to ensure their reliability, due to the vast amount of noise that is generated in any given application. Modern computer processing power and the availability of signal processing software will enable reliable systems to be supplied.

10.6 Reliability-centred maintenance (RCM)

Reliability-centred maintenance analysis is a procedure for determining maintenance strategy. It was originally developed for the airline industry due to the high cost of planned maintenance. It was shown that planned maintenance resulted in some 80 per cent of items being replaced that showed no signs of age-related deterioration.

The application of RCM will need to be selective for other industries due to the different needs and operating cycles when compared with airlines. The method of analysis is useful and will show some benefit if applied in sections of operation where there is a need to modernize and apply a mix of maintenance strategies.

10.6.1 Procedure
The procedure requires a review of the total plant or machine so that all failure modes are identified. In this, it will be helpful to subdivide the plant into production units or the machine into subassemblies. Once all the failure modes are identified, then the consequences for each failure

can be defined. This will enable the failures to be ranked in accordance to their impact on safety and cost. Cost could be lost output and/or high cost of repair. Failure of a machine or plant requires a maintenance response for it to be returned to operation. This requires manpower and material resources. The aim of RCM is the optimum balance of all these factors. The procedure can be formalized with the steps shown in Table 10.3.

Table 10.3 Steps in RCM analysis

	Steps required	*Action needed*
1	System definition	Acquisition of data on the operating and reliability requirements, develop block diagrams for analysis.
2	Identify the maintenance-significant items	Using FTA and Pareto analysis as needed. Find the items whose failure will significantly threaten safety, or increase cost due to lost production or have a high cost of repair.
3	Identify the failure modes	Using FMEA. Find the causes of failure and how they could be detected.
4	Select the maintenance strategy	Using the RCM decision process. For each failure mode decide what can be done to reduce its likelihood of occurrance or to mitigate its consequences.

	Implementation
1	The formation of a task list into a workable plant-wide schedule with organizational responsibilities, manpower loading and material requirements.
2	Implementation of the work schedule with sustained feedback of in-service data for periodic review and update.

Note

Pareto, an Italian engineer and statistician, showed that in a multiple of tasks there are only a small minority that have the most effect. The task is to identify them, as this gives the maximum return for the least effort.

10.6.2 RCM decision process

The RCM decision process is in two steps: the first is to categorize the failure modes and the second is to select the appropriate maintenance strategy.

Failure mode categories

It is important to sort failure modes into categories. These categories can be ranked by their consequences (see Table 10.4) and this will give guidance on the action needed.

Table 10.4 Consequence analysis

	Category	Description
1	Hidden failure	Not detected during normal operation but affects safety and/or reliability. Applies to non-operating standby equipment and non-fail-safe protective equipment.
2	Safety/environmental consequences	Failures that cause loss of function or secondary damage which could have a direct impact on safety or the environment.
3	Operational consequences	Failures that have a direct adverse effect on operational capability.
4	Non-operational consequences	Failures that do not affect operations, for example where there are installed redundancies.

Task analysis

This ranks the maintenance options in order of preference, see Table 10.5.

Table 10.5 Task analysis

	Maintenance strategy	Basis for selection
1	Condition-based maintenance	If the ability to detect potential failure is applicable and worthwhile.
2	Restoration	If the possibility of repair to reduce failure rate is applicable and worthwhile.
3	Replacement	If the possibility of replacement to reduce failure rate is applicable and worthwhile.
4	Combination	If the possibility of a combination of maintenance strategies to reduce failure rate is applicable and worthwhile.
5	Redesign	If none of the above are acceptable, further investigation and analysis may be required and the whole design concept may need to be reconsidered.

In each case it is important to question. Is the task under consideration both applicable and worthwhile? Could it be done? Would it work? Would its cost, direct and indirect, be less than just allowing the failure to occur?

This underlines the fact that the lowest direct maintenance cost is the use of breakdown maintenance. The equipment is used until it wears out and no operating cost is involved in monitoring its condition. It is the possible consequences that preclude its use. This stresses the need to categorize the failures.

The procedure is carried out for each category of failure mode in turn. In many cases the selection process will be speeded up, based on experience and established practice.

10.6.3 RCM application

The RCM procedure, having been developed and successfully applied for the airline industry, may not be useful for every situation. For the aeronautical industry it was successful because there were a large number of the same models being produced for different airlines, but operating in similar conditions. Aeroplanes, in general, use common parts and systems, which helps in the gathering of generic data and the provision of good information on failure rates. Other industries may not be so fortunate. Operating requirements differ from industry to industry and these differences may dominate the selection of a maintenance strategy.

- Power generating plant may be subject to consumer demand. Some plants only operate on peak demand and others are affected by seasonal demand.
- Process plants need to shut down to meet legal requirements for inspection.
- Pressure vessels and boilers are subject to the need for regulatory inspections.
- Some plants manufacture to stock and, when sufficient buffer is available, shut down.
- LNG liquefaction plants supply customers by the use of LNG tankers. Demand can be seasonal and dependent on tanker maintenance and survey requirements. There is excess storage capacity available to allow for shipping delays and other contingencies.
- Designed storage and redundancy feature in the plant design.
- Plants which only work one or two shifts and shut down for the weekend.

These operating circumstances will demand maximum reliability between scheduled shut-downs. Engineers will very often be required to do everything possible to keep things going until the scheduled shut-down. For these situations the RCM process will need some adjustment.

10.7 Summary

It has been suggested that plant and machinery should be designed for a service life with zero maintenance. This goal has yet to be achieved for most industrial plant.

Maintenance engineering remains a critical factor in ensuring safety and reliability. The technology is under continuous development for the reduction of operating costs. However, it is also being increasingly recognized as a profit centre. Any improvement in plant reliability is also reflected in improved output. This is a business benefit, being a gain in cash flow and an improved return in capital investment.

With all the technological advances, however, research has shown that reliable operation is still dependent on the skill and care of the operators and maintenance engineers. People can notice signs and symptoms and form conclusions that make a difference. This has been demonstrated where plant and machinery of the same design, but operated at two different locations with different operators, will have different failure rates.

Chapter 11

Piper Alpha

11.1 Introduction

A study of the events that led to the Piper Alpha disaster (1) will serve to illustrate all the issues discussed in the preceding chapters of this book. Piper Alpha was the name of an oil and gas production platform situated in the North Sea about 340 km east of Aberdeen in Scotland. The platform was mounted on a steel structural support, called a jacket, resting on the seabed that was some 140 m deep. Oil production started in December 1976. Later, gas was also exported in 1978. Figure 11.1 shows Piper Alpha in production.

In July 1988 there was an explosion, and fire broke out which destroyed the platform with the loss of 166 lives. This disaster was a turning point in the law with regard to safety. As a result of the Cullen inquiry into the disaster, it was concluded that a complete change in the law was needed (1). Piper Alpha complied with all the safety regulations current at the time but these did not save it from disaster. As a result, the law was changed and now, in addition to being prescriptive, requires safety objectives to be met.

Fig. 11.1 Piper Alpha in production

11.1.1 Operating requirements

Piper Alpha was designed to produce crude oil. In the production of crude oil some associated gas is produced and this waste gas was burnt in a flare where the flame is discharged into the atmosphere. The oil field was found to be very productive and the operating company wanted to increase production. As the UK government regulated production, permission was granted on condition that the gas would be processed and transmitted to the mainland for distribution by British Gas. This requirement resulted in the need for gas processing facilities that were not catered for in the original design. As the platform area was limited, the new gas processing facilities could only be accommodated with the control and communications centre, together with the electrical distribution centre, placed above them. This then resulted in the accommodation module being placed as another layer above the control room level, with the helicopter landing deck on top. The processing arrangement is shown in Fig. 11.2.

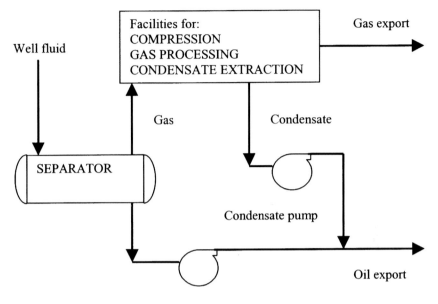

Fig. 11.2 Piper Alpha oil and gas processing

11.2 Export arrangements

A sub-sea pipeline to the Flotta on-shore terminal exported the oil produced by Piper Alpha. Two nearby platforms, named Claymore and Tartan, were also producing oil and gas. The produced crude was pumped into the same pipeline to Flotta, being connected to a T-junction downstream from Piper Alpha. A sub-sea gas pipeline to the MCO-01 platform, however, transmitted the produced gas where it was discharged into the pipeline from Frigg field, to St Fergus on-shore gas terminal. The produced gas from the nearby Claymore and Tartan platforms was also sent to MCO-01, but via Piper Alpha. How these platforms were interconnected is shown in Fig. 11.3.

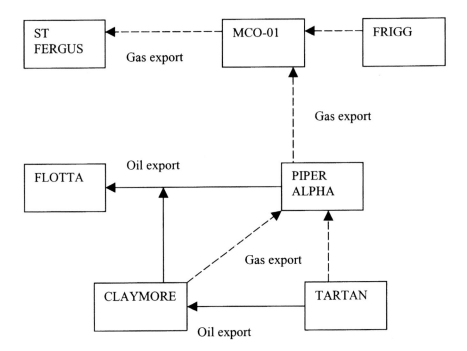

Fig. 11.3 Piper Alpha oil and gas import/export arrangements

11.3 The disaster

The disaster happened very quickly when it started on 6th July 1988 and very soon most of the crew were dead. The casualties were as follows:

Complement	226 men
Survived	61
Died	165
In addition, rescuers killed	2

Cause of death:	
Smoke inhalation	109
Drowning	13
Severe injuries and burns	10
Burns and infection	1
Missing	30

All the management died and only one control room operator survived. The events of the disaster had to be pieced together, see Table 11.1.

Table 11.1 Event log

Date	Time	Event
6th July 1988	21.45	Condensate pump trip alarm in control room
	21.50	As observed in the control room: Gas alarm in gas processing area First-stage gas compressor trip alarm Waste gas flare seemed larger than usual
	22.00	The first explosion occurred The oil and gas separation area and the oil export pump area on fire; ESD operated Accommodation module engulfed in smoke
	22.20	Due to the heat from the fire, the high-pressure gas line connecting Tartan to Piper Alpha exploded
	22.40	Tartan shut down
	22.50	The high-pressure gas export pipeline to MCO-01 exploded
	23.00	Claymore shut down
	23.20	The final high-pressure gas pipeline, which connected Claymore, exploded The heat of the fire was so intense the topsides structure was being weakened and started to fall into the sea; one part that fell was the accommodation module with 81 men inside
7th July 1988	Early morning	Most of the topsides and sections of the jacket had collapsed; only the well head module was left
29th July 1988		Fires extinguished
28th March 1989		The remains of Piper Alpha toppled into the sea

It was later calculated that the fractured gas pipes were each discharging gas initially at a rate of 3 t/s with gas flames consuming up to possibly 100 GW and reaching a peak height of some 200 m.

Figure 11.4 shows Piper Alpha on fire and Fig. 11.5 shows Piper Alpha destroyed.

Fig. 11.4 Piper Alpha on fire

Fig. 11.5 Piper Alpha destroyed

11.4 The reconstruction of events

As with most disasters, the incident was caused by a combination of events that was fatal.

Maintenance operations

Evening of 6th July 1988

The condensate pump, which injected condensate into the crude oil export line, had an installed spare to provide 100 per cent redundancy, see Fig. 11.6. This allowed maintenance work to be carried out without disrupting production. That night, pump A was shut down and isolated for maintenance of its motor drive coupling. Opportunity was also taken to remove its PRV for maintenance. A blank flange was fitted in its place to cover the opening, as was the normal practice. The blank flange covering the hole was not leak or pressure tested. It was placed there to keep the pipe clean, as is normal good practice, and so it was certainly not pressure tested. It was very likely that only a few bolts with finger-tight nuts were fitted to keep it in place.

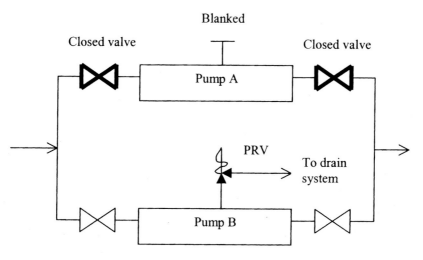

Fig. 11.6 Condensate pump arrangement

Night of 6th July, 21.45

Production was normal but for some reason condensate pump B tripped. The operators tried to start it a number of times and each time it tripped out.

The whole production output of the platform depended on running a condensate pump. That was the reason for installing a spare pump. If the condensate was not removed, then the level in the separator before the inlet to the final-stage compressor would reach danger. There would be an alarm and the plant would shut down.

The operators were aware that pump A was isolated and shut down for maintenance. The permit system was in operation but there was no mention that the PRV was removed for maintenance. The pump was shut down for routine maintenance of the motor drive coupling, which was all they knew.

Manning
The night shift consisted of:

- the operations superintendent;
- the deputy operations superintendent;
- the lead production operator;
- two well-head area operators;
- two gas process area operators;
- a control room operator.

Conjecture on the explosion
Because of the information available to them, it is likely that the operators would see no reason for not putting pump A back into operation. As far as they were aware, it was down for maintenance of the motor drive coupling. The coupling was still in place and so the work had not started. Unfortunately the PRV, contrary to normal practice, was located in the floor above. This was due to the need to ensure proper drainage facilities. The fact that the PRV was missing could not be seen, and there was no reason for the operators to look.

The operators' duty was to maintain production, and so it is highly probable that they decided to run pump A.

On opening up the valves and repressurizing the pump, it is fairly certain that condensate would have been discharged from the loose blanking flange. It has been estimated that possibly some 90 kg could have been discharged in about 30 s. It is very possible that this was the source of the first explosion.

Other related events

Firewater pumps
The firewater system autostart was turned off and manual control was selected. At the time of the disaster, the jacket legs were scheduled for

underwater inspection. There was concern that, should a pump be started, a diver could be sucked in at a pump intake and suffer some injury. This was in spite of the fact that the firewater pump had grills to protect the intakes. Unfortunately the pump manual starters were located near the fire and in spite of valiant efforts they could not be reached.

ESD valves

The ESD valves on the crude oil export lines did not close oil tight. The initial explosion in the export pump area also fractured the export oil pipes. Tartan and Claymore continued production for almost an hour after the initial explosion until they shut down. The sub-sea export pipeline continued in use during this time and some crude oil was also being fed into the fire on Piper Alpha via the leaking ESD valve. This leakage flooded the area and also ran down to the deck below; where the gas import/export pipelines terminated. This fire then led to the ultimate fracture of the gas pipelines and the final destruction of Piper Alpha.

Evacuation order

Neither the off-shore installations manager nor his deputy ever issued the order to abandon the platform. They were the only persons authorized to do so. The 61 men who survived abandoned the platform in defiance of standing orders. Other men stayed on the platform, thinking that they would be rescued by helicopter. No liferafts or lifeboats were successfully launched.

Helicopter rescue

At the time, 226 helicopters were available for rescue operations. This was impossible as the landing pad was engulfed by smoke almost immediately.

Communications

The control room and the radio room were put out of action within 20 min of the first explosion. No signals or messages were sent to the other interconnected platforms in that time. This accounted for the time delay in shutting down Tartan and Claymore. If Tartan and Claymore had shut down within minutes of the first explosion, it is possible that the scale of the disaster could have been reduced.

Work permit

Because the motor drive coupling had not been removed, it was decided that the work permit would not be posted until the morning maintenance shift came on duty. The work permit was not posted and sat in the safety office. Pump A, however, remained isolated ready for maintenance. It would appear that the situation was blurred. The fact that the PRV had been removed did not seem to be accounted for.

Isolation

There were no security isolation facilities used. The pump switchgear was racked out, but there was no locking procedure and so anyone could just rack it back in. The normal procedure for isolation was to attach an isolation warning tag. Although isolation of hazardous gas was required, just single isolation valves were used, with nothing to prevent them being opened. They were pneumatic-operated valves and the air supplies were disconnected, but it was an easy matter to reconnect them with local actuator control to cause them to open. Security of isolation, therefore, just relied on warning tags, with no other deterrent.

Risk management

No formal risk management procedures were in place other than the work permit system. However, in addition to plans for evacuation by helicopter, a multi-function support vessel was in place. This was the support ship *Tharos*. The *Tharos* was close by and available to be of assistance to Piper Alpha throughout the disaster but was impotent. It had significant fire fighting capability and when they witnessed the explosion they immediately came alongside to help fight the resulting fire. Unfortunately in the excitement, just by chance, all the firewater pumps were switched on at the same time and the ship suffered a power failure. After power had been restored, because all of the fire monitors had been left open the firewater main was not at the correct pressure and so the firewater pumps could not operate. Valuable time was lost. This lost time, and the fact that the fire was escalating by being fed with fuel, meant that the fire fighting efforts of the *Tharos* had no effect.

The final reckoning

1. 166 men died;
2. 10 per cent of UK oil production lost;
3. £2000 million financial loss (1988 value).

11.5 Comments

This case study serves to illustrate the various points made throughout the book. These can be highlighted as follows.

Complacency

This is the most common of all mistakes to make and has been the cause of many disasters. There had never been a fire and so people thought that there could never be one. Hazards must have been considered in design and there must have been good reasons for the installation of all safety features. If there is a compelling reason for disabling any safety feature, then some contingency plan must be in place to counter any hazard that might arise. The crew disabled the automatic fire protection system to safeguard the divers but no thought was given as to what to do in the event of a fire. This shows that any change will increase risk and that a full safety case has to be prepared and authority obtained to ensure safety is not compromised, as required by the management of HSW regulations.

Hazards of change

The change in function of Piper Alpha meant the need to get a quart into a pint pot. It was designed to produce crude oil and was changed to increase output and at the same time produce export gas.

These changes restricted the design with regard to the location of hazards and the ability to arrange plant in the safest way. The design met all the applicable regulations at the time. It really demonstrated that they were not enough and that the laws had to be changed. This again demonstrates how any change in function or design will increase risk, and that this must be managed.

The reliability of ESD valves

The ESD valve that did not close oil tight contributed to the escalation of the fire. Unreliable ESD valves are not acceptable. One outcome of the disaster has been a concerted effort on the development of more reliable ESD valves and ESD systems. Fireproof ESD valves are now available, tested to be operable, and capable of tight shut-off even in a fire.

The work permit system

The case study underlines the importance of the need to:

1. change jurisdiction for maintenance operations;
2. control the scope of work;
3. secure isolation;
4. formally hand over at shift changes;
5. ensure effective communication;
6. ensure effective risk management.

Risk management

The incident illustrated the importance of emergency planning and training. As demonstrated, when an incident occurs there needs to be a completely different mind-set to prevent escalation. The first thought of the disaster management team would have been to think of how to reduce casualties. This will be the order to abandon the platform. How to do it and how much time was available for evacuation would need to dominate their minds. This will be in addition to how to protect the remaining assets.

Safety case

The Off-shore Installations (Safety Case) Regulations SI (1992) No. 2885 now require operators to submit to the HSE a safety case which must demonstrate that safety objectives, which can be verified by independent persons, have been met. This is of importance, as this approach will be increasingly applied where there is a public concern for safety. The requirements for a safety case will include and demonstrate that:

1. the safety management of the company is adequate to ensure a safe design and safe operation of the installation;
2. all potential hazards have been identified and sufficient action has been taken to control the risks;
3. adequate emergency planning and training is in place and a temporary safe refuge is provided for, with adequate rescue and evacuation provisions made.

11.6 Summary

It is hoped that this case study has proved to be a suitable ending for this book. All the various important issues that have been expounded will have been illustrated by this study.

It has shown that safety and reliability can be inextricably linked. Reliable production was lost as a result of the lack of safety. It has also shown that safety and reliability can only be achieved by the joint efforts of engineering, operations and maintenance. The duty of management is to ensure the integration of the work of all three.

11.7 Reference

(1) **Cullen, Lord** (1990) *The Public Enquiry into the Piper Alpha Disaster*, HMSO, London, ISBN 010113102X.

Glossary

ALARP	As low as reasonably practicable
API	American Petroleum Institute
ASME	American Society of Mechanical Engineers
ATEX	Atmosphere, explosive
BASEEFA	British Approvals Service (for Equipment in Flammable Atmospheres)
CCTV	Closed circuit television
CEN	European Committee for Normalization (Standards)
CENELEC	European Committee for Electrotechnical Standardization
CO_2	Carbon dioxide
DCC	Disaster control centre
DIN	German National Standards (translated from German)
EC	European Community
EEC	European Economic Community
EECS	Electrical Equipment Certification Service as provided by BASEEFA
ESD	Emergency shut-down
EU	European Union
FMEA	Failure mode and effects analysis
FTA	Fault tree analysis
HAZOP	Hazards and operability (studies)
HSE	Health and Safety Executive
HVAC	Heating, ventilation and air conditioning
ICAO	International Civil Aeronautics Organization
IEC	International Electrotechnical Commission
IES	Illuminating Engineering Society
IMO	International Maritime Organization
IP	Institute of Petroleum
ISO	International Standards Organization

LMTD	Log mean temperature difference
MTTF	Mean time to fail
MTTR	Mean time to repair
NDT	Non-destructive testing (ultrasonic, radiography, etc.)
NEC	National Electrical Code (USA)
NEMA	National Electrical Manufacturers Association (USA)
NOX	Nitrous oxide
NR	Noise rating as defined by ISO
OSHA	Occupational Safety and Health Administration (USA)
PAHH	High pressure alarm and trip
PC	Pressure control
PFD	Process flow diagram
P&ID	Piping and instrument diagram
PRV	Pressure relief valve
QA	Quality assurance
QC	Quality control
RCM	Reliability-centred maintenance
SOLAS	Safety of Life at Sea Regulations as issued by IMO
TESEO	Technica Empirica Stima Errori Operati (Italy)
TUV	Technical Surveillance Association (translated from German; an international network of test laboratories)

Bibliography

Andrews, J.D. and **Moss, T.R.** (1993) *Reliability and risk assessment*, Longman Scientific, Harlow, ISBN 0 582 09615 4.

Bloch, H.P. and **Geitner, F.K.** (1990) *Machinery reliability assessment*, Van Nostrand Reinhold, New York, ISBN 0 442 23279 9.

Burrough, B. (1999) *Dragonfly, NASA and the crisis aboard MIR*, Fourth Estate, London, ISBN 1 84115 087 8.

Canning, J. and **Ridley, J.** (1998) *Safety at Work*, Butterworth-Heinemann, ISBN 0 7506 4018 9.

Carter, A.D.S. (1986) *Mechanical reliability*, Second edition, Macmillan, London, ISBN 0 333 40587 0.

Carter, A.D.S. (1997) *Mechanical reliability and design*, Macmillan, London, ISBN 0 333 69465 1.

Davidson, J. and **Hunsley, C.** (Eds) (1994) *The reliability of mechanical systems*, Second edition, Mechanical Engineering Publications, IMechE, London, ISBN 0 85298 881 8.

The Engineering and Technology Board. *Guidelines on risk issues.*

Health and Safety Executive, UK. *Free Guides*, as listed in their books catalogue.

IMechE Seminar 2000-1, *Reliability of sealing systems for rotating machinery*, P.E.Publishing, ISBN 1 86058 245 1.

Kirwan, B. and **Ainsworth, L.K.** (1992) *A guide to task analysis*, Taylor and Francis, ISBN 0 7484 0057 5.

Kletz, T. (1992) *Hazop and Hazan*, IChemEng, ISBN 0 85295 285 6.

Kletz, T. (2001) *What went wrong? Case studies of process plant disasters*, Butterworth-Heinemann, ISBN 0 8841 5920 5.

Kletz, T.A. (1991) *An engineer's view of human error*, Second edition, IChemEng, ISBN 0 85295 265 1.

Lees, F.P. (1996) *Loss prevention in the process industries*, Butterworth-Heinemann, Oxford, ISBN 0 7506 1547 8.

Moubray, J. (1991) *Reliability-centred maintenance*, Butterworth-Heinemann, Oxford, ISBN 0 7506 0230 9.

Occupational Health and Safety Administration, USA.
Guides to OSHA standards, available on their website.

O'Connor, P.D.T. (1995) *Practical reliability engineering*, Third edition, (Revised), John Wiley, Chichester, ISBN 0 471 96025 X.

Petroski, H. (1985) *To engineer is human, the role of failure in successful design*, St Martin's Press, New York, ISBN 0 312 80680 9.

Wong, W. (1997) *Process machinery – safety and reliability*, P.E.Publishing, ISBN 1 86058 046 7.

Vance, J.M. (1988) *Rotordynamics of turbomachinery*, John Wiley, ISBN 0 471 80258 1.

Helpful HSE guides (a selection)

HSG48 *Reducing error and influencing behaviour.*

HSG65 *Successful health and safety management.*

HSG85 *Electricity at work (safe working practices).*

HSG138 *Techniques to reduce noise in the workplace.*

HSG142 *Dealing with off-shore emergencies.*

HSG176 *The storage of flammable liquids in tanks.*

HSG181 *Assessment principles for off-shore safety case.*
(NB Could be generally useful.)

HSG Books catalogue (for many more titles).

Informative BSI/ISO standards

BS 4778-32: 1991 *Quality vocabulary.*

BS 5760-0: 1986 *Reliability of systems, equipment and components.*
Introductory guide to reliability.

BSI PDI 5304 *Safe use of machinery.*

ISO IEC 60300 *Dependability management, Part 1 management systems, Part 2 guidance for dependability management.*

ISO IEC 61025 *Fault Tree Analysis (FTA).*

ISO IEC 61078 *Analysis techniques for dependability, reliability block diagram methods.*

ISO IEC 61160 *Formal design review.*

ISO IEC 61882 *Hazard and operability studies (HAZOP studies) – Application guide*, First edition.

Videos

(Produced by the Mechanical Reliability Committee, IMechE, and obtainable – cost on application – from the Continuing Education Unit, University of Manchester School of Engineering, The University, Oxford Road, Manchester M13 9PL, UK.)

Dissecting system failures. Fault tree analysis, technique and computerized application.

Exploring failure consequences. Failure mode and effect analysis, technique and applications.

Learning from failures. Weibull reliability analysis, technique and applications.

Directory

Government agencies

European Commission, 8 Storeys Gate, SW1. *For information on EEC Council Directives on Health and Safety*, see websites www.cec.org.uk and http://europa.eu.int/comm/employment_social/h&s/index_en.htm

Health and Safety Executive, Rose Court, 2 Southwark Bridge, London, SE1 9HS. *For information on UK Laws and Regulations on Health and Safety*, see www.hse.gov.uk. Also see www.hsedirect.com

H.M. The Stationery Office, 119 Kingsway, London, WC2. *The full text for all UK legislation, including those on Health and Safety, can be found on their website* www.hmso.gov.uk

Occupational Health and Safety Administration, USA, www.osha.gov *For addresses of local offices and the download of OSHA guides to standards.*

US Coast Guard. *Reports on the human element in safety.* See reports and studies and the Human Element Bibliography Resource Page. www.uscg.mil/humanelements

Professional bodies

American Institute of Chemical Engineers www.aiche.org

American Petroleum Institute www.api.org

American Society of Mechanical Engineers www.asme.org

American Society of Safety Engineers www.ASSE.org

European Safety Reliability and Data Association, Beeckzanglaan lc, 1942 LS Beverwijk, The Netherlands. www.vtt.fi/aut/tau/network/esreda/esr_home.htm

Institute of Petroleum www.petroleum.co.uk

Institution of Chemical Engineers, Davis Building, 165-171 Railway Terrace, Rugby, Warwickshire, CV21 3HQ. http://icheme.chemeng.ed.ac.uk

Institution of Mechanical Engineers www.imeche.org.uk

Safety and Reliability Society, Clayton House, 59 Piccadilly, Manchester, M1 2AQ. www.sars.u-net.com

Safety Engineering and Risk Analysis Division (SERAD), ASME International. www.asme.org/divisions/serad

National organizations

British Safety Council, 70 Chancellors Road, London, W6 9RS.

The Engineering and Technology Board, 10 Maltravers Street, London, WC2. www.engc.org.uk

Fire Protection Association UK www.thefpa.co.uk

The Loss Prevention Council, Building Research Establishment, Bucknells Lane, Garston, Watford, WD25 9XX. www.bre.co.uk

Royal Society for the Prevention of Accidents, Edgbaston Park, 353 Bristol Road, Edgbaston, Birmingham, B5 7ST. www.rospa.co.uk

Certifying and inspection agencies

BASEEFA (EECS). *UK certifying authority for electrical and mechanical equipment and protective systems used in flammable atmospheres, and for other electrical safety related certifying requirements. Also known as EECS, has mutual recognition agreement with Factory Mutual, USA.* www.baseefa.com

Bureau Veritas www.Bureauveritas.com

Lloyd's Register www.lr.org

TUV. *EU notified body for testing and certification of equipment to comply with EU directives. Has worldwide offices and has a memorandum of understanding with Underwriters Laboratories, USA.* TUV technical services, www.tuvglobal.com and TUV product services, www.tuv.com

Consultants

Advantage Technical Consulting, The Barbican East Street, Farnham, Surrey, GU9 7TB. www.advantage-business.co.uk

AEA Technology, Risley, Warrington, WA3 6AT, UK. www.aeat.co.uk

AMEY Vectra, 310 Europa Boulevard, Gemini Business Park, Westbrook, Warrington, Cheshire, WA5 5YQ. www.ameyvectra.co.uk

BMT Reliability Consultants Ltd, Fernside, 12 Little Farm Road, Farnham, UK. www.bmtrcl.com

DNV Technica Ltd, Palace House, 3 Cathedral Street, London, SE1 9DE, UK. www.dnv.com

Envirocare Technical Consultancy. *For all health and safety matters.* http://home.clara.net/envirocare

Eutech, Belasis Hall Technology Park, PO Box 99, Billingham, Cleveland, TS23 4YS, UK. www.eutech.co.uk

P M Safety Consultants Ltd, The Verdin Exchange, High Street, Winsford, Cheshire, CW7 2AN. www.pmsc.u-net.com

Primatech. *Consultants and services for all OSHA requirements.* www.primatech.com

Quest Consultants Inc. *Consultants and services for all OSHA requirements.* www.questconsult.com

R M Consultants Ltd. *Risk management consultants and application software.* www.rmcnorth.demon.co.uk

Education and training

Institute of Risk Research, Vienna, Austria. www.irf.univie.ac.at

Primatech. *Consultants and services for all OSHA requirements.* www.primatech.com

Quest Consultants Inc. *Consultants and services for all OSHA requirements.* www.questconsult.com

School of Industrial and Manufacturing Science, Cranfield University, Cranfield, Bedford, MK43 0AL. *Safety, risk and reliability modules.* www.cranfield.ac.uk

SERCO Assurance, Thomson House, Risley, Warrington, Cheshire, WA3 6AT. *Safety and risk management training.* www.sercoassurance.com

University of Loughborough. *Courses on safety and reliability.* Loughborough, Leicestershire, LE11 3TU.

University of Manchester. *Maintenance/reliability engineering IGDS courses.* School of Engineering, Division of Mechanical Engineering, The Simon Building, University of Manchester, Oxford Road, Manchester, M13 9PL.

Fire protection engineering services and equipment suppliers

Air Sense Technology Ltd, www.airsense.co.uk

AMEC, www.amec.co.uk

Chubb Fire Systems, Security House, Fiveways Business Centre, Aspen, Feltham, TW13 7AQ. www.chubb.co.uk

Electro-Detectors, www.electrodetectors.co.uk

EMS Group, www.emsgroup.co.uk

Hi Fog, Marioff Corporation Oy, PO Box 25, FIN 01511 Vantaa, Finland. *Watermist fire protection systems*, www.hi-fog.com

Tyco/Fire & Security (Wormald Fire Systems), Wormald Park, Grimshaw Lane, Newton Heath, Manchester, M40 2WL. www.wormald.co.uk

Tyco International. *Fire protection and fire detection equipment and services*, www.tycoint.com

Instrumentation

ADT Fire and Security, Security House, The Summit, Hanworth Road, Sudbury on Thames, UK.

Bently Nevada (UK) Ltd, 2 Kelvin Close, Science Park, Birchwood, Warrington, Cheshire, WA3 7BL. *Machinery condition monitoring.* www.bently.com

Bruel and Kjaer. *Sound and vibration monitoring.* www.bksv.com

entek IRD Mechanalysis (UK) Ltd, Bumpers Lane, Sealand Industrial Estate, Chester, CH1 4LT. *Machinery condition monitoring.* www.entekird.com

Honeywell SA, Bourgetlaan 3, 1140 Brussels, Belgium. *Plant monitoring and control.* www.iac.honeywell.com

Itronics. *UV, IR, and UV–IR optical flame detectors.* Via Applegate directory www.applegate.co.uk

Malin Instruments. *Diesel engine monitoring.* www.malin.co.uk

Zellweger Analytics Ltd. *Gas and vapour detection.* www.zellweger-analytics.co.uk. Or via Applegate directory www.applegate.co.uk

Noise control

Acoustic Associates, www.acousticassociates.co.uk

Acoustics Profile, www.noice.co.uk

Applied Acoustics Design, www.aad.co.uk

Envirocare Technical Consultancy, *for all health and safety matters, including noise.* http://home.clara.net/envirocare

ISVR Consultancy Services, www.isvr.co.uk

Safety system suppliers

Castell Safety International Ltd, Kingsbury Road, London, NW9 8UR. *Integrated safety solutions, interlocks.* www.castell.com

Giro Engineering Ltd, Talisman, Duncan Road, Parkgate, Southampton, Hants, SO31 7GA. *Sheathed diesel fuel injection pipes.*
www.giro.dial.pipex.com

Loadtec Engineered Systems Ltd, The Stables, Coach House, Hythe Road, Smeeth, Kent, TN25 6SP, UK. *Safety systems for bulk liquid transfer.* www.loadtec.co.uk

Tyco Valves and Controls, www.tycovalves.com

Software

Advantage Technical Consulting, The Barbican East Street, Farnham, Surrey, GU9 7TB. *Risk modelling simulation software.*
www.advantage-business.co.uk

BMT Reliability Consultants Ltd, Fernside, 12 Little Farm Road, Farnham, UK. *Risk analysis software.* www.bmtrcl.com

Dyadem International Ltd. *A complete range of software.*
www.dyadem.com

IsographDirect. *Risk analysis software.* www.isographdirect.com

Item Software (UK) Ltd. *A complete range of software.* www.itemuk.com

Primatech, *Process hazard analysis (HAZOP etc.) software.*
www.primatech.com

Quest Consultants Inc. *Incident-modelling software.*
www.questconsult.com

Safety and Reliability Society. *Their website provides information on software and links to providers. See 'Search the site'.*

System for Production Availability and Resources Consumption (SPARC). www.program42-ies.com

Data sources

EIReDA, European Industry Data Handbook, Editions SFER, Paris.

OREDA (1992) Offshore Reliability Data Handbook. Det Norske Veritas Industri, Norge AS, DNV Technica.

RELDAT™ from AEA Consultancy Services, Risley, Warrington, UK.

Index

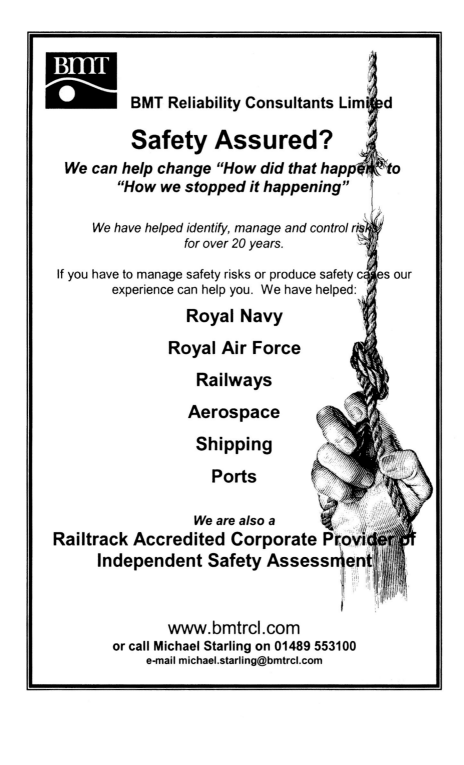

CCPS Is Your Safe Bet

When it comes to process safety, you simply can't afford to gamble. That's why CCPS gives you the resources you need to maximize safety — and your resources. No other single source offers the quantity and quality of process safety tools available from CCPS. Odds are, we have exactly what you need. So bet on CCPS as *the* source for safety.

To become a CCPS Sponsor...to purchase ProSmart...to purchase books or register for courses...or to participate in CCPS collaborative projects:

Visit our Web site at www.aiche.org/ccps;
e-mail ccps@aiche.org;
call 212.591.7319; or fax 212.591.8895

CCPS process safety

educational courses at convenient locations

Over 70 books available on process safety

The 2002 Annual Conference and Workshop, *Risk and Reliability,* Jacksonville, FL, October 8-11, 2002

ProSmart: Performance measurement of PSM systems

Collaborative Industry Projects
• Process Equipment Reliability Database
• Process Safety Incident Database
• Safety and Chemical Engineering Education

CCPS
CENTER FOR
CHEMICAL PROCESS SAFETY
An AIChE Industry
Technology Alliance

© 2001 American Institute of Chemical Engineers